Revaluing Construction

Written and Edited by

Peter Barrett

Published in association with CIB
The International Council for Research and Innovation
in Building and Construction

Editorial offices:
Blackwell Publishing Ltd, 9600 Garsington Road, Oxford OX4 2DQ, UK
  Tel: +44 (0) 1865 776868
Blackwell Publishing Inc., 350 Main Street, Malden, MA 02148-5020, USA
  Tel: +1 781 388 8250
Blackwell Publishing Asia Pty Ltd, 550 Swanston Street, Carlton, Victoria 3053, Australia
  Tel: +61 (0) 3 8359 1011

First published 2008 by Blackwell Publishing Ltd

Library of Congress Cataloging-in-Publication Data:

Barrett, Peter, professor.
  Revaluing construction / written and edited by Peter Barrett.
    p.   cm.
  Includes bibliographical references and index.
  ISBN-13: 978-1-4051-5919-7 (hardback : alk. paper)
  ISBN-10: 1-4051-5919-7 (hardback : alk. paper)
  1. Building—Quality control.   2. Construction industry—Customer services.   3. Construction industry—Management.   I. Title

TH438.2.B37   2007
690.068′5—dc22
2007006852

ISBN 978-1-4051-5919-7

A catalogue record for this title is available from the British Library

Set in 9.5/12 pt Palatino
by Aptara Inc., New Delhi, India
Printed and bound in Singapore
by Utopia Press Pte Ltd

The publisher's policy is to use permanent paper from mills that operate a sustainable forestry policy, and which has been manufactured from pulp processed using acid-free and elementary chlorine-free practices. Furthermore, the publisher ensures that the text paper and cover board used have met acceptable environmental accreditation standards.

For further information on Blackwell Publishing, visit our website:
www.blackwellpublishing.com/construction

CIB's mission is to serve its members through encouraging and facilitating international cooperation and information exchange in building and construction research and innovation. CIB is engaged in the scientific, technical, economic and social domains related to building and construction, supporting improvements in the building process and the performance of the built environment.

CIB Membership offers:
- international networking between academia, R&D organisations and industry
- participation in local and international CIB conferences, symposia and seminars
- CIB special publications and conference proceedings
- R&D collaboration

Membership: CIB currently numbers over 400 members originating in some 70 countries, with very different backgrounds: major public or semi-public organisations, research institutes, universities and technical schools, documentation centres, firms, contractors, etc. CIB members include most of the major national laboratories and leading universities around the world in building and construction.

Working Commissions and Task Groups: CIB Members participate in over 50 Working Commissions and Task Groups, undertaking collaborative R&D activities organised around:

- construction materials and technologies
- indoor environment
- design of buildings and of the built environment
- organisation, management and economics
- legal and procurement practices

Networking: The CIB provides a platform for academia, R&D organisations and industry to network together, as well as a network to decision makers, government institution and other building and construction institutions and organisations. The CIB network is respected for its thought-leadership, information and knowledge.

The CIB has formal and informal relationships with, amongst others: the United Nations Environmental Programme (UNEP); the European Commission; the European Network of Building Research Institutes (ENBRI); the International Initiative for Sustainable Built Environment (iiSBE); the International Organization for Standardization (ISO); the International Labour Organization (ILO); International Energy Agency (IEA); International Associations of Civil Engineering, including ECCS, fib, IABSE, IASS and RILEM.

Conferences, Symposia and Seminars: CIB conferences and co-sponsored conferences cover a wide range of areas of interest to its Members, and attract more than 5000 participants worldwide per year.

Publications: The CIB produces a wide range of special publications, conference proceedings, etc., most of which are available to CIB Members via the CIB home pages. The CIB network also provides access to the publications of its more than 400 Members.

Themes: The main thrust of CIB activities takes place through a network of around 50 Working Commissions and Task Groups, organised around three CIB Priority Themes:

- Sustainable Construction
- Performance Based Building
- Revaluing Construction

A fourth priority Theme, Integrated Design Solutions is currently being developed within CIB.

Contents

Contributors   vii

Preface   x

Acknowledgements   xii

PART 1   INTRODUCTION/OVERVIEW   1

1   Introduction/need for change   3
     Peter Barrett and Angela Lee

2   Overview of agenda for change   7
     Peter Barrett

PART 2   SEVEN PRIORITY AREAS   9

3   Holistic idea of construction   11
     Peter Barrett

4   Shared vision amongst stakeholders   18
     Peter Barrett

5   Balance of markets and social capital   30
     Peter Barrett

6   Dynamic decisions and information   36
     Peter Barrett

7   Evolving knowledge and attitudes   46
     Peter Barrett

8   Awareness of systemic contribution   51
     Peter Barrett

9   Promotion of full value delivered by society   58
     Peter Barrett

PART 3   VIEWS AROUND PRIORITY AREAS   65

10   Assessing the true value of construction and the built environment   67
     Les Ruddock

11   Competing revaluing construction paradigms in practice   83
     George Ang

12   The trajectory of construction procurement in the UK   105
     Chris Goodier, Robby Soetanto, Andrew Fleming, Peter McDermott
     and Simon Austin

13   Delivering full value through seamless information systems   115
     Ghassan Aouad, Nick Bakis, Song Wu and Emeka Efe Osaji

14   Long-term educational implications of revaluing construction   130
     Melvyn A. Lees

15   Revaluing construction: a building users' perspective   149
     Jacqueline Vischer

16   Construction is good!   164
     Angela Lee and Peter Barrett

17   A wider view: revaluing construction in developing countries   174
     George Ofori

PART 4   IMPLICATIONS IN PRACTICE AND CONCLUSIONS   193

18   Exemplars of 'revalued' construction   195
     Lucinda Barrett

19   Stakeholder action areas   209
     Peter Barrett

20   Conclusions   218
     Peter Barrett

APPENDICES   223

A1   Members of the CIB Revaluing Construction Steering Panel   225

A2   Sequence of activities and sources feeding into the CIB revaluing
     construction proactive theme   226

A3   Sample revaluing construction covering letter and survey
     questionnaire   234

A4   International survey questionnaire results   240

     References   261

     Index   279

Contributors

Professor Peter Barrett
Professor Barrett is the Pro Vice Chancellor for Research & Graduate Studies at the University of Salford, Chairman of Salford Centre for Research and Innovation, and, as of 2007, President elect of the CIB. To date, he has produced over 100 single volume publications, refereed papers and reports, and has delivered over 110 presentations in around 16 countries. He has undertaken a wide range of research, typified by management focus around real world problems.

George Ang
George Ang is an architect and civil engineer by profession. He has practised architectural design in mainly educational facilities, and was highly specialised within that field. He has lead on development and implementation of large scale design and build for the Dutch Government Building Agency. Currently (since 2004) he is assigned with strategic implementation management of change for the PPP and is an active board member of the PSIBouw.

Professor Ghassan Aouad
Professor Aouad is Dean of the Faculty of Business, Law & the Built Environment at the University of Salford, and Director of the £5m EPSRC IMRC Centre. He led the prestigious £443k EPSRC platform grant (from 3D to nD modelling). His research interests are in: modelling and visualisation; development of information standards; process mapping and improvement; and virtual organisations.

Professor Simon Austin
Professor Austin is currently Professor of Structural Engineering in the Department of Civil and Building Engineering at Loughborough University. He has undertaken industry-focused research for over 25 years into design processes, modelling, integrated working and management techniques, information management, process re-engineering, value management and structural materials and their design. He is the author of over 200 publications and holder of over 20 EPSRC-funded research grants.

Nick Bakis
Mr Nick Bakis is a Research Fellow at the Salford Centre for Research and Innovation (SCRI) at the University of Salford. He holds an MSc and BSc in Computer Science and has been involved in the area of Construction IT over the last ten years.

Lucinda Barrett
Lucinda Barrett is a researcher in Salford University's Centre for Research and Innovation in the built environment. She has an MA in English Studies from the University of Manchester. She has worked on the Revaluing Construction project and Leadership

Foundation project as a facilitator in the workshops both nationally and internationally. She is currently coordinating an international workshop on the design of space.

Andrew Fleming

Andrew Fleming is currently based at the University of Salford where he has conducted construction sector research since 1998 in the areas of process management, change management and competitiveness.

Dr Chris Goodier

Dr Goodier began his career at Laing Civil Engineering, working on major civils' projects such as the M5 and Second Severn Crossing. After completing his PhD he worked for four years for BRE in concrete consultancy before returning to Loughborough as a senior researcher in construction futures, offsite and modern methods.

Dr Angela Lee

Dr Lee is the Programme Director for the BSc (Hons) in Architectural Design and Technology course at the University of Salford. She is/has worked on various EU-, CIB- and EPSRC-funded projects. Her research interests include: design management; performance measurement; process management; nD modelling; product and process modelling; and requirements' capture. She has published extensively in both journal and construction papers in these fields.

Professor Melvyn A. Lees

Professor Lees' current appointment is as Professor of Quantity Surveying and Education in the School of the Built Environment at the University of Salford. He is currently Head of School and Deputy Director of the Centre for Education in the Built Environment. He teaches postgraduate courses in project construction and his main research interests are in education supply chains.

Professor Peter McDermott

Dr McDermott is a professor in management at the School of the Built Environment in the University of Salford. He is a founder member and was Joint Co-ordinator of the CIB in 1990 (International Building Research Council) Working Commission W92 – Construction Procurement Systems. Dr McDermott is Editor of the *International Journal of Construction Procurement*. He has published widely in the field of construction procurement and acts in an advisory capacity to a range of public and private sector organisations on this topic.

Dr George Ofori

Dr Ofori is a professor and Head of Department of Building at the National University of Singapore. His research area is construction industry development, focusing on developing countries. He is Co-ordinator of CIB W107 on Construction in Developing Countries and consultant to international agencies and governments. Dr Ofori has authored some 200 papers, books and reports.

Emeka Efe Osaji

Emeka Efe Osaji is an architect with over five years' experience in the built environment, and will conclude his PhD research study at the University of Salford in 2007. He has worked as a Research Assistant with Professor Ghassan Aouad, and as a Research

Assistant on the Construction and Black Minority Ethnic (BME) research project. His research interests include: architectural morphology; energy performance of the spheroid form; sustainable and bioclimatic design; and innovative design in the built healing environment, amongst others. Emeka Efe Osaji has recently been appointed a Research Associate at Loughborough University and will work on the HaCIRIC research project with Professor Andrew Price.

Professor Les Ruddock

Professor Ruddock is Professor of Construction and Property Economics and Associate Dean for Research in the Faculty of Business, Law and the Built Environment at the University of Salford. He is Co-ordinator of the CIB Task Group on Macroeconomics for Construction, and has written extensively on the roles of the construction and built environment sectors within the wider economy.

Dr Robby Soetanto

Dr Soetanto's educational background is in civil engineering and construction management. Before taking up his academic career, he worked in the property sector. He has been involved in several industry and government-sponsored research projects. His main research interest concerns human-related factors in various built environment contexts, and he has published widely in this area.

Dr Jacqueline Vischer

Dr Vischer is an environmental psychologist who has specialised in the environmental psychology of the work environment. She has published some six books and numerous articles on this subject. Her most recent book is *Space Meets Status: Designing Workplace Performance*. She is Professor of Interior Design at the University of Montreal, where she directs the New Work Environments Research Group.

Dr Song Wu

Dr Wu is a lecturer of Quantity Surveying at the School of Built Environment, University of Salford. His research interests are in: modelling and visualisation, data modelling and simulation, and software engineering.

Preface

Revaluing construction is a growing theme with worldwide reach that is providing a focus for the question: how can the construction industry maximise its contribution to the creation of value within society?

This is a very broad perspective that opens up questions about the fundamental conception taken of the industry. Of course an efficient, cost-effective industry is desirable, but the revaluing construction view shifts attention to include issues concerning the holistic role of the built environment. It asks questions about what are good environments for learning, getting well, working efficiently, performing creatively, living comfortably, etc. And then, from this understanding, asks how can the industry take on a service-orientated attitude and capacity that can deliver these optimal built environment settings. In addressing these questions the industry itself has the potential to be very positively reassessed within society.

This book sets out seven interlinked areas through which, it is argued, this reframing of the industry can be achieved. These will impact on how the industry is seen and, just as importantly, how it sees itself. The seven areas make up a meta-analysis of a multitude of factors, which highlights what seem to be key leverage points and connects them in a progressive flow. Working from a holistic, use-driven, view of the built environment, the value of an ambitious vision for construction amongst all key stakeholders is stressed, including the equitable distribution of the rewards associated with the value created. From this the question of the optimal balance between markets and relationships (social capital) is addressed, and this links into issues of dynamic decisions and information that should flow progressively from conception through to use. To move beyond isolated improvements the knowledge and attitudes of all involved in construction, including clients, have to be enhanced. If all this can be achieved the industry will operate at a much higher level of alignment with society's needs. For this to be credited to the industry it is important that the systemic contribution of the built environment to practically every part of our lives is better understood and, further, that this contribution is actively promoted.

In addition to a focused exposition of the above ideas, a range of international experts have each taken one of the seven areas and provided an extended contribution from their specialist perspective. The book finishes with examples of 'revalued' construction and practical actions for the various stakeholders. Using evidence from international experience, it is stressed that some things can be achieved quite quickly, but that for significant and comprehensive improvement a concerted effort on all of the areas highlighted will be required for a decade or so.

This book will be of interest to policy-makers in many countries. For example, built environment activity in the UK makes up 20% of GDP, thus to improve the efficiency and effectiveness of this sector in itself and as a foundation for all other sectors is incredibly important. Similarly, clients and procurers of construction services will find the book of value as it highlights both the degree to which they can and should demand physical settings tuned to their needs. For the construction industry this book clarifies many of the obstacles to progress and raises a wide range of opportunities for moving past these to achieve enhanced performance. For the academic and educational community it is hoped that the ideas, directions and connections highlighted here will provide inspiration and impetus to future work that will collectively support industry change agenda.

Peter Barrett

Acknowledgements

Many people have contributed to this project and I would like to thank them all.

Particular thanks must go to the participants and respondents to the surveys and workshops carried out in America, Australia, Canada, Singapore and the UK. I would like to take this opportunity to thank the 60 workshop attendees for enthusiastically taking part and giving willingly of their time and intellects. In each location we were crucially supported by a local co-organiser: Professor Colin Davidson in Montreal, Dr Florence Ling in Singapore, Professor Fred Moavenzadeh in Boston and Professor Tony Sidwell in Brisbane.

Five very valuable mini-reports in specialist areas were written in support of the project by: Dr George Ang of the Government Building Agency in The Netherlands; Professor George Ofori of the University of Singapore; Professors Mel Lees and Les Ruddock of the University of Salford; and Professor Jacqueline Vischer of the University of Montreal. These mini-reports have been developed into chapters and supplemented by chapter contributions from Professor Ghassan Aouad's University of Salford team and the cross-university 'Big Ideas' project team.

I had the pleasure of being involved in an investigation of practices in Norway and Denmark for the Dutch Government and would like to thank Dr George Ang for this opportunity, which proved very timely.

The project had a very knowledgeable Steering Panel drawn from the coordinators of the relevant CIB Working Commissions and Task Group. The members are listed in Appendix 1 and their wisdom is greatly appreciated.

Without all this input, some funding from the CIB and the support of the Research Institute of the Built and Human Environment and SCRI, both at Salford University, it would not have been possible to carry out the project. I am very grateful for this support and the friendly and supportive reaction that has invariably been forthcoming.

Special mention must be made of the high level of commitment and support that Dr Angela Lee has provided throughout the project, and particularly in relation to the postal survey and the creation of an excellent web site. This was complemented by wise and insightful contributions from Lucinda Barrett, especially during the series of workshops, out of which she created the 4Cs model. In the latter stages of production Anna Higson's relentless pursuit of final details was crucial.

I hope the report properly reflects the views of all those who have contributed and brings out strong, shared messages that will be of interest to all involved. Of course any errors and mistakes remain the responsibility of the author/editor.

Peter Barrett
University of Salford, UK

Part 1
Introduction/overview

1 Introduction/need for change

Peter Barrett and Angela Lee

Introduction

Construction is often seen as an embattled industry. For example, the UK construction industry has been under sustained and increasing pressure to improve its practices (Howell, 1999; Smith et al., 2001). Since the 1940s, it has been continuously criticised for its less than optimal performance by several government and institutional reports, such as Simon (1944), Emmerson (1962), Banwell (1964), Latham (1994), Egan (1998) and, more recently, Fairclough (2002). Most of these reports conclude, time and time again, that the fragmented nature of the industry, lack of co-ordination and communication between parties, the informal and unstructured learning process, adversarial contractual relationships and lack of customer focus are what inhibit the industry's performance. In the words of one report: '... there is a deep concern that the industry as a whole is under-achieving' (Egan, 1998). Construction projects are also often seen as unpredictable in terms of delivery time, cost, profitability and quality, and, in addition, that investment into research and development is usually seen as expensive when compared to other industries (Egan, 1998; Fairclough, 2002).

Pearce's (2003, page ix) recent report for nCRISP, the UK's Construction Research and Innovation Strategy Panel that provides leading recommendations to the Government, summarises these factors and contends on the industry's self-image problem:

> ... those within the industry, and some outside it, criticise what they see as poor economic performance relative to other countries. They point to problems of adapting to rapid technological change, and to the highly skewed size structure of the industry with many thousands of small firms inhibiting the capture of economies of scale. They worry about the social image of the industry's workforce, about the health and safety record of the industry, and about the skills structures and international competitiveness.

These factors are typical of the construction industry in general. In Australia, demands made in similar reports by NPWC/NBCC (1990), Gyles (1992), NSW (1992), CIDA (1995) and DIST (1999) for a more efficient and effective industry have paralleled the analogous concerns to those in the UK. More too, in Ireland the Barry Report (Barry, 1997) made recommendations on the 'internal' operation of the industry, including contracts, tendering procedures, dispute resolution, etc.

The repeated critique of all these reports thus questions the ability of the construction industry to innovate and manage change to improve its practices (Lansley, 1987; Gale &

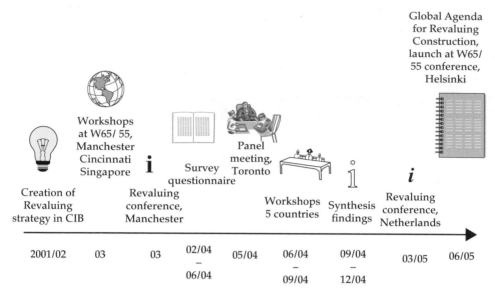

Figure 1.1   Summary programme of revaluing construction theme within CIB.

Fellows, 1990; Betts & Ofori, 1993; Barrett, 2002). According to Howell (1999), the 'inefficiency' of the construction industry has tended to be the way of life. This may be due to the fact that none of the reports have been significantly acted upon. As Latham (1994) points out:

> . . . some of the recommendations of the reports were implemented. . . but other problems persisted, and to this day, even the structure of the industry and nature of many of its clients has not changed dramatically.

So, is change in the industry's make-up plausible or even appropriate to bring about widespread improvement/innovation?

The priority theme for revaluing construction was initially settled upon within the International Council for Research and Innovation in Building and Construction (CIB) in 1997 at the CIB Board meeting in South Africa. At this stage it was styled as 're-engineering construction'. However, it did not really get off the ground until 2001 when work was put in hand to create a strategy for the development of the theme. This was carried out by Professors Courtney and Winch and led to a re-orientation around the notion of 'revaluing construction' and a stream of activities that is summarised in Appendix 2. A simple overview is given in Figure 1.1.

Another key shift in the direction of the initiative came out of the discussions stimulated by the initial strategy, and this was to work to engage the CIB Working Commissions/Task Groups in the development of the theme. There is a wealth of relevant knowledge within the CIB membership, as indicated by Figure 1.2.

The issues involved are complex, as indicated by the mapping given in Figure 1.3.

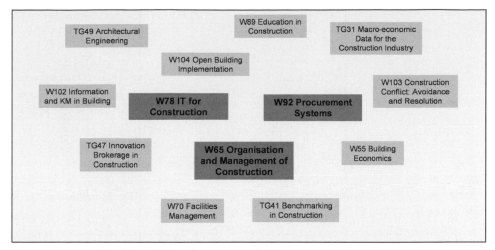

Figure 1.2 Wealth of knowledge from CIB WCs/TGs.

Figure 1.3 represents the outcome from five workshops (Barrett & Barrett, 2004), each in a different country, namely: Australia, Canada, Singapore, UK and USA. These provided the main basis for the focal 'infinity' model that underpins the organisation of the material in this book. A postal questionnaire survey to the five countries (Barrett & Lee, 2004) also provides pivotal underpinning and is given in Appendix 3. Further, five specialist reports were commissioned and informed the analysis, and have since been developed into chapters. The main objective remained, however, to identify the significant points of leverage to make progress without over-simplifying a patently complex situation.

The book is in four parts:

- **Part 1,** comprising Chapters 1–2, briefly introduces the issue and then summarises the overall agenda for change. This then shapes the rest of the book.
- **Part 2,** comprises Chapters 3–9, which each pick up one of the seven identified areas for action and set out the arguments for the importance of the area.
- **Part 3,** made up of Chapters 10–17, returns to each of the seven areas and provides an in-depth specialist view on an important aspect of the area.
- **Part 4,** consists of Chapters 18–20, which respectively provide exemplars of 're-valued' construction, suggested stakeholders/actions and, finally, conclusions.

It is hoped that the areas identified and the actions illustrated, together will lead to a situation where the views of the construction industry cited above will rapidly become a thing of the past. Instead the crucial role of construction in society will be optimally delivered and fully appreciated.

There is a lot of rich material to follow and so, before plunging into this, the next chapter seeks to set out, in simple summary form, the overall arguments that are central to this book.

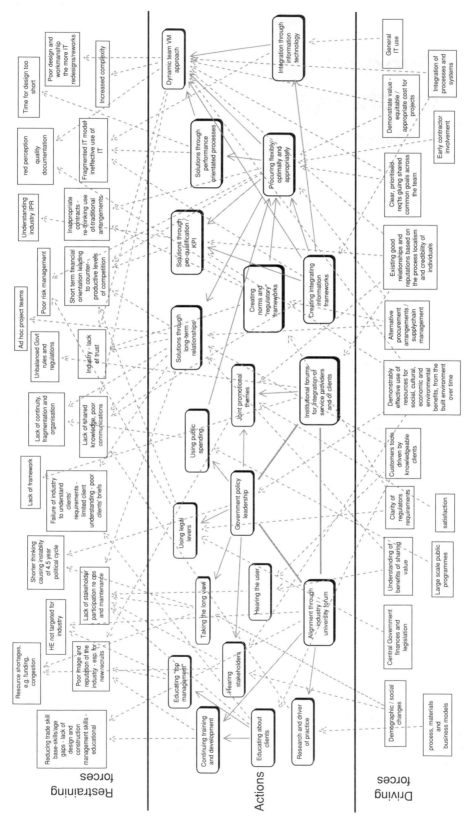

Figure 1.3 Synthesis map drawn from all five workshops.

2 Overview of agenda for change

Peter Barrett

The main thrust of results of this work on revaluing construction is encapsulated in seven areas for change. Each of these is not in itself extraordinary, but when dynamically linked have the potential to fundamentally change construction for the good of those involved, their customers and for society. The 'infinity' model given in Figure 2.1 outlines these seven areas and suggests how they must be connected to make progress. At this stage a summary of the territory set out by this perspective will be given. Subsequent sections will expand on each of the seven areas.

The top-left box in the figure will be mentioned first to stress the importance of the basic conception taken of the notion of construction. It is pivotal to revaluing construction that a broad, holistic idea is adopted, otherwise the potential of the industry to maximise its contribution to buildings in use will be compromised. From this robust basis the creation of a shared vision amongst stakeholders (at the centre of the figure) can be addressed that emphasises maximising the value jointly created and equitably distributing the resulting rewards. This political consensus creating process is primarily located at a national policy level involving major stakeholders. It is here that the vision for revaluing construction is created, maintained and promulgated, including its practical implications. Within this conducive policy context, a key operational area where significant change is needed is in the balance of weighting between market forces and social capital, particularly in relation to procurement. When appropriately addressed, to provide a higher level of stability and trust, there appears to be significant willingness to handle information and decisions more coherently and dynamically throughout the whole building life-cycle. This then has the potential to release considerable latent gains in value. Taken together, these actions will mean that some clients and some projects will deliver much higher levels of value. However, to make the improvements take hold across the industry in the longer term it is essential that the knowledge and attitudes of those involved evolve strongly. This will then reinforce isolated good practice so that it becomes normal practice. The three boxes on the right-hand half of the figure, together with their interactive connection to the central vision, provide a clear focus on how the industry can move to improve its performance by 'looking in' at the practices, relationships and techniques that it employs.

In itself this will deliver great benefits. However, it will be relatively fragile and in a sense will not seriously shift the limited and often negative perception of construction within society. For the role of construction to be significantly revalued the industry needs to 'look outwards' and work to raise awareness of the systemic contribution

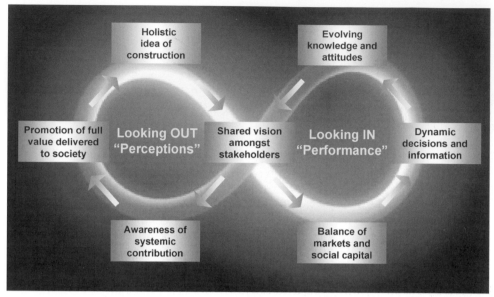

Figure 2.1   Global agenda for revaluing construction.

that construction makes. This is indicated in the bottom left-hand box in the figure and involves accounting for the multiple value streams running from construction, some for very many years beyond the building event itself. Given the generally negative standing of the industry, the final step is to pick up on the rich and positive messages contained in the contributions identified and to actively promote the full value delivered to society by construction. Success on this front will then bring us back to the box that started this description, by reinforcing the holistic idea of construction not only as a compelling theoretical idea but as a powerful policy and social conception as well.

The 'infinity' model stresses the two complementary halves of the revaluing construction agenda: the industry looking in at itself to perform better, and also looking out at how it is perceived within society. The seven action areas and their connections are proposed as a coherent set of priority areas that taken together provide a dynamic improvement process for the industry as a whole. The 'infinity' model illustrates and symbolises the ongoing, reinforcing nature of the path set out. What is described in this report is a journey rather than a destination. On the way, though, construction will progressively raise its contribution and profile within society.

The following sections discuss each of the seven areas and expand on the issues, challenges and opportunities.

Part 2
Seven Priority Areas

3 Holistic idea of construction

Peter Barrett

Introduction

There are starkly different ways of conceiving of construction. The Standard Industrial Classification (SIC) system is the basis of the normal economic perspective and places construction within F45, a cluster of activities that includes: site preparation; building of completed constructions, or parts thereof; civil engineering; building installation; building completion; and renting of construction or demolition equipment with an operator. The focus is entirely on the physical construction activities. It draws a line between these and intimately linked, value-adding activities, such as even the parallel work of architectural and technical consultancies. Moreover, upstream activities – such as the design activities of consultancies, related business services and the significant construction dimensions of manufacturing, mining and quarrying – are omitted from the definition of construction. Less surprisingly, downstream activities – such as real estate activities and facilities management concerned with the use phase of buildings – are also left out (Ruddock & Wharton, 2004).

A holistic economic view

An alternative stance starts with the proposition that construction is a change agent for the creation, development and maintenance of the built environment so that it supports the quality-of-life and competitiveness requirements of society. That is, 'construction is a means to a means to an end' (Barrett, 2003). This makes a broader conception of construction entirely logical so that its full contribution to society can be understood. This type of thinking is central to the work of Jean Carassus of CSTB in Paris, who with a group of international colleagues developed and populated a shared economic framework within Task Group 31 of the CIB (Carassus, 2004). From the original notion in 1999 and the creation of the group in 2000, the collaboration took three years and involved testing the utility of the framework through case studies from nine diverse countries: Australia, Canada, Denmark, France, Germany, Lithuania, Portugal, Sweden and the UK. Their resulting coordinating model, given in Figure 3.1, focuses on a 'meso-economic' analysis of the construction sector. Carassus distinguishes this from either micro-economics (with its focus on individuals and firms buying, selling, setting prices and making profits) or macro-economics (with its emphasis on average prices, employment and production, taxes and national government policy). Meso-economics does not replace either view, but suggests an additional complementary perspective

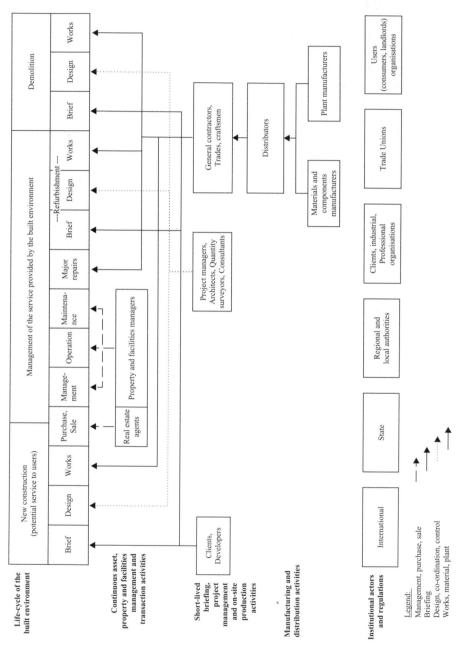

Figure 3.1 TG 31s meso-economic framework for the construction sector system.

that places the spotlight on a specific sector of the economy in its entirety, and so highlights issues such as industry structure, political issues of economic development and policy, regional issues and the influence of groupings and associations.

The framework takes the full building life-cycle (from left to right) of: new construction; management of the service provided by the built environment; and demolition. Included in the middle phase are: maintenance; major repairs; and refurbishment. Taking a vertical view of the framework (from top to bottom) leads from the stream of activities required to create and sustain the built environment, to the panoply of actors or stakeholders with varying degrees of involvement. This ranges from real estate agents and property and facilities managers with an ongoing involvement with the asset, to those with a short-lived involvement via projects, such as developers, project managers, architects and contractors. Underpinning their activities are the associated contributions of manufacturers and distributors. Lastly, contextualising the whole sector are the institutional actors at various geographical levels together with professional representative organisations and user associations themselves. These together infuse the sector's norms, regulations and expectations. The thrust of Carassus et al. is to shift thinking from an 'industry' focus on simply building buildings, to a 'construction sector system' approach with the emphasis on producing and managing the *services* rendered by these structures throughout their life-cycle to support an efficient and sustainable economy.

The implications running from this richer perspective are significant. Under the SIC classification, construction typically amounts to around 6% added value share in GDP. For very developed countries there is a slight downward trend and an upward trend for other countries. However, taking the construction sector to include an appropriate share of the contributing industries typically doubles the added value figures to over 12% of GDP. This excludes the value of the facilities management dimension of the use phase, but figures collated by Gerard de Valence for Australia (see Carassus, 2004) show that including this aspect brings the total closer to 20% of GDP. As well as identifying the huge contribution of the sector, this broader perspective also highlights the importance of the existing stock and its maintenance, adaptation and management, together with the significant role played by small companies, usually carrying out small projects. At a policy level it is not very common to stress these aspects and so a 'sector' perspective will facilitate a move away from solutions for just the big, public and new projects.

Chapter 10 examines these economic issues in more detail, providing international data within a sustainable development perspective. Arguments are also given for action to achieve more consistent, holistic data sets so that policy and practice are not inadvertently confounded.

Broader implications

The above perspective also brings into focus the importance of the necessary interactions between the numerous stakeholders so that joint progress can be made. For example, these include: designers seeking new materials; facilities managers and emerging needs for different ways of using buildings; manufacturers developing competitive products and services; public authorities promoting environmentally improved

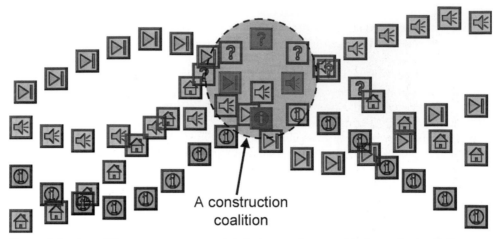

Figure 3.2   A coalition as a temporary state between streams of independent activity.

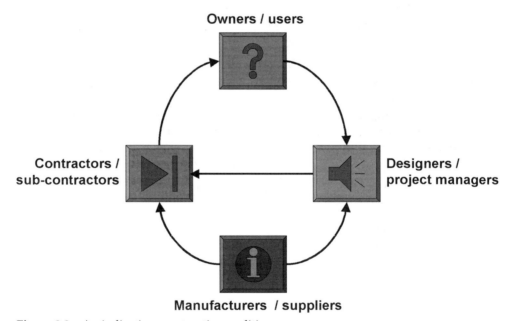

Figure 3.3   An indicative construction coalition.

practices; and educational institutions working to support a sustainable and competitive industry. These interactions are dynamic and complex. Taking 'the project' as a focus, Figures 3.2 and 3.3 illustrate the notion of a number of independent companies, whose turnover and trajectory comprise a stream of multiple projects, that come together at a particular point in time to form a temporary 'coalition' around a specific project (Fellows et al., 1983).

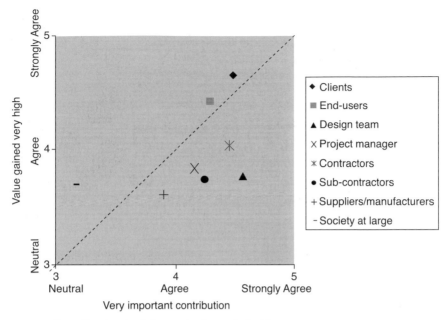

Figure 3.4 Contributing/gaining value by stakeholder.

This project then draws and combines resources from the firms, after which they rejoin their individual streams of different projects. The grouping formed is termed 'a coalition' to stress that those involved bring distinctive contributions, work together, but retain distinctive ends rooted in their longer term trajectories. Interestingly, Nicolas (1999) defines a sector as a cluster of firms, sharing a similar paradigm, that together evolve appropriate processes and structures. In the terminology of this report these would be sub-sectors within which particular players interact, using particular approaches around distinctive types of work. Linking these ideas it is evident that the construction sector system is a much broader conception than 'construction', but that it holds within it the necessity to distinguish distinctive sub-clusterings if the particular sets of relationships are to be fully understood. Using the material in Carassus's (2004) work, initial sub-sectors can be suggested, such as: civil engineering; housing; non-residential buildings; repair; and maintenance. Within these, coherent layers of the meso-economic framework could fruitfully be studied.

It is important to seek ways in which these distinctive clusters can work together to maximise the value created; but to be sustainable, those involved must benefit as well as simply contribute. A postal survey of five countries[1] (Lee & Barrett, 2006) addressed this issue, as shown in Figure 3.4.

This shows a combination of two questions asking, for each stakeholder, respectively the contribution made and the value gained. The diagonal line gives the equilibrium where contribution and benefit would be in broad balance. It is evident that the 'construction team', and in particular the design team, feel they contribute at a higher level

---

[1] Australia, Canada, Singapore, UK, USA

than the benefit they derive. Clients and end-users appear to be in broad equilibrium, and 'society at large' is perceived by the respondents to gain more than it contributes. This last is to be expected given that 'society' generally ensures its requirements are met through regulation. It should be noted that the rating scales used extended in the negative direction but the aggregate responses are solidly in the positive quadrant (shown). This is an indication of a good level of mutual recognition. However, overall there remains the feeling that the industry senses it is not fully rewarded. Of course companies get financial returns for their involvement, which can be crucial to short-term survival and profitability. On a softer dimension, rewards around achievement and kudos should not be overlooked. Everybody thinks it was 'their' project, particularly when it is a success! This is a great source of satisfaction and can become critical to longer term success as it builds into a high reputation for the company and individuals and good relationships with other players. Underpinning this should ideally be the transfer and progressive accumulation of knowledge and experience from projects to the firms themselves. This again reinforces the importance of working in specific sub-sectors or clusters where competence can be created and traded. This was made very clear in presentations at the Revaluing Construction Conference in Rotterdam in 2005 when senior directors of HEVO and Haskell independently highlighted the importance

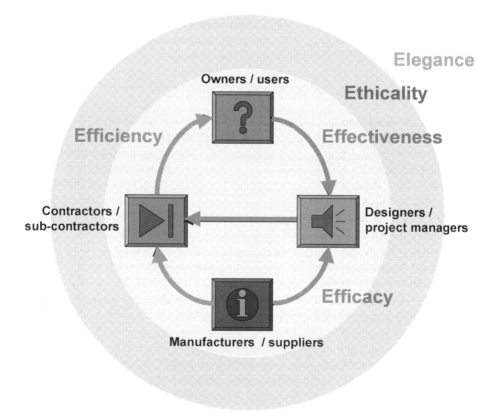

Figure 3.5   Performance and the construction coalition.

of garnering specialist knowledge and constraining their operations to defined 'market segments'.

A range of stakeholders have been indicated, but is interesting to consider the relevant performance criteria for the relationships between the parties. Drawn from 'soft systems' the holistic 5Es typology (Checkland & Scholes, 1990) can be summarised as:

Efficacy: did the process produce the required output?
Efficiency: was the transformation carried out with minimum resources?
Effectiveness: were long-term requirements met?
Ethicality: was the transformation acceptable from a value judgement perspective?
Elegance: was the solution over- or under-engineered, aesthetically pleasing, well designed?

Mapping these criteria onto the coalition diagram gives Figure 3.5. Various distinctive characteristics for the various relationships can be identified. Effectiveness is central to the owner/designer/contractor relationship. The designers require efficacy in their relationship with manufacturers/suppliers who in turn have a relationship with contractors typified by efficiency, which runs through the delivery of the built solution to owners. This slight extension of the previous discussion reinforces the complexity and specificity of the relationships involved that will evolve to suit specific clusters of players in coherent and distinct sub-sectors of the overall construction sector system.

Summary

In summary, an initial step towards revaluing construction is to adopt a *holistic idea of construction* that is orientated towards the primary objective of creating and sustaining appropriate built environments for users. This whole life-cycle view includes *more* than is often encompassed by economic definitions and the normal building aspects, and extends to: clients; designers; contractors; specialist sub-contractors; manufacturers; suppliers; and facilities managers. It also covers: small as well as big projects/players; the private sector as well as the public sector; repair and refurbishment as well as new build. It is important to link this broadening of perspective to focusing on *less* for specific analyses by accepting that the industry is in fact a bundle of sub-industries. Within these, clusters of players and issues combine in specific and particular ways. In short, a holistic systems overview is needed, but it should be linked to focused contingency analyses for naturally distinguishable sub-sectors. A significant academic effort is necessary to move from a position of competing, seemingly incommensurable, paradigms to a rich, but robust consensus that can support powerful progress.

This chapter has begun to identify the full range of stakeholders connected to construction. The next chapter will consider how they can move to fashion a shared vision for revaluing construction within the broad context provided here.

4 Shared vision amongst stakeholders

Peter Barrett

Introduction

The broader conception of construction set out in Chapter 3 can inform the creation of a shared vision amongst key stakeholders for maximising value across the whole life-cycle of constructed artefacts. Figure 4.1 portrays the 'building life-cycle', including the usual project/building orientated phases of design – construction – use, but then extending in scope to a societal or urban scale and linking back to the start through the construction industry as a key part of society. At each stage there is a focus, drawn from lean thinking on maximising value-adding activities and minimising non-value-adding activities. This is indicated by the 'up' and 'down' arrows.

Taking each phase in turn, the first concerns building 'potential' value through a design process, where the emphasis has to be heavily on maximising value-adding activities and not trying to short-cut matters to save time, avoid difficult debates, etc. The next phase focuses on delivering maximum value through construction, that is maintaining as much of the potential value as possible, implying an emphasis on minimising non-value-adding activities that routinely drain away the enthusiasm of all participants. The third phase is where the building is put to use and the value is actually realised. Here, it is suggested, there is a need for both maximising value-adding and minimising non-value-adding activities so that the potential envisioned in the design phase is communicated to and inspires users, but at the same time the impediments to the inevitable adaptations needed are removed so that the users and building evolve in concert over time. The following phase takes the consideration beyond the individual project and building and raises the question as to how synergistically the cumulative impacts of individual developments add up at the urban level. Here the emphasis is likely to be on maximising the impact of value-adding activities, many at the early planning stages, but also at the creative, operational end as well. Within this societal context, the construction industry is the key change agent for the built environment (Barrett, 2003). A revalued industry will maximise the initial creation of *potential* value in a particular building/project through pre-design and design activities, its *delivery* through construction, *realisation* in use and *synergies* with other developments at an urban level. The various dimensions of this overview are discussed in detail in the body of this book, but the desired impacts on value through the building life-cycle are shown diagrammatically in Figure 4.2.

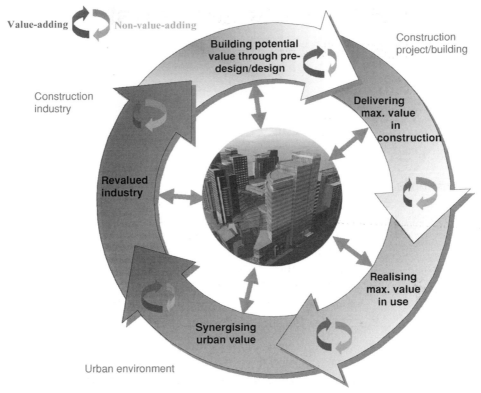

Figure 4.1 Building value cycle.

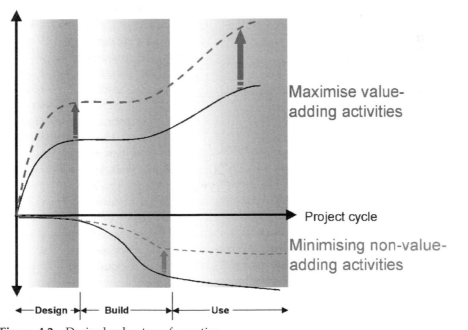

Figure 4.2 Desired value transformation.

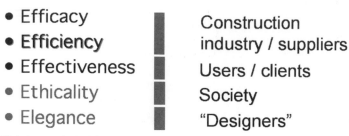

- Efficacy
- **Efficiency**
- Effectiveness
- Ethicality
- Elegance

Construction
industry / suppliers
Users / clients
Society
"Designers"

Figure 4.3   Driving stakeholders for 5Es.

Stakeholder involvement

To inform the wide-ranging scope of this conception of construction it is clear that a broad constituency of stakeholders needs to be engaged to develop and sustain a vision for construction. In the five-country questionnaire (see Appendices 3 and 4) and workshops the following stakeholders were involved: clients and users; designers; contractors and sub-contractors; government regulators; and manufacturers and suppliers.

In fact, the last group was very difficult to engage despite considerable efforts, and there is some concern that this indicates that manufacturers and suppliers do not really consider themselves to be an intrinsic part of the construction industry. This in itself deserves attention given the significant role these companies perform and the research and development resources that they command.

It has been suggested in Chapter 3 on a 'holistic idea of construction' that the various stakeholders have linkages that impact on performance against the 5Es criteria. This is summarised in Figure 4.3 and highlights the likely limitations of a vision produced by just the industry. Without the manufacturers and suppliers involved many opportunities for improved efficacy and efficiency would not surface; without active involvement from clients and users' effectiveness is likely to be inadequately addressed; and without societal representation in terms of planners and representative groups the ethicality of proposals is likely to be under-developed. Lastly, without some design input the competing demands of the multitudinous and diverse stakeholders are unlikely to be resolved into elegant solutions. On this last point, two aspects need clarification, first, the term 'elegant' is meant conceptually and practically in that the solutions involve sufficient complexity, but are as simple as they can be. Second, 'designers' here are not building designers, but rather 'leaders as designers' (Senge, 1990, pp. 341–345) at various levels who understand the complexities of the 'system' and put together the ideas and the infrastructure in ways that enable the industry to perform to its full potential. At the top level come shared visions held jointly by key national, leading, stakeholders.

Bringing together the relevant groups and engendering productive co-working is a complex political process. Within the UK this has been in train progressively over the last ten years. The main construction-related Government department was central to the initiative and first encouraged the formation of broader groupings of existing representative groups. For example, the Construction Industry Council is a combination of the representative bodies of architects (RIBA), surveyors (RICS), contractors (CIOB)

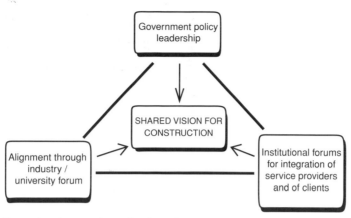

Figure 4.4　Focus for the creation of a shared vision.

and others, which themselves are substantial, but fairly numerous players. Once a certain amount of consolidation had occurred on this and other fronts, the Government focused attention by liasing primarily with these broader groupings and establishing the Construction Research and Innovation Strategy Panel (CRISP) as a forum within which these broader groupings themselves could meet, with the primary remit of creating an industry level strategy. Each country will need to take an approach that fits their historical and cultural characteristics. Interestingly, it was revealed through the study visits for the Dutch Government's industry reform programme (PSIBouw, 2004) that there is a long-standing collaboration in Norway between the professional institutions associated with construction, owing in part to their co-location in the same building. The same series of visits garnered information about the Danish Association of Construction Clients that was encouraged by their Government to grow in an incremental and very lean form. As with the UK some central seed-corn funding was used in combination with active dialogue, so legitimising and giving value to the existence of the grouping. Once established, these focused national groups can benefit from connections with like groupings from elsewhere. The establishment by the CIB of TG58 focused on clients and innovation, and the VROM/CIB series of client conferences (VROM, 2004; CIB, 2004) are both examples of international initiatives to support this particularly important stakeholder.

The form of the Danish clients' association raises an issue that moves the discussion onto a different level. In the UK the creation of CRISP was seen as the solution, with a single forum within which the main players in the industry could generate a strategic view. The Danish clients' forum is strongly independent, and linking this point to the wide range of stakeholders with something to offer, as indicated earlier in this section, the notion of a dialectic between various stakeholder voices seems a looser, but arguably more dynamic and richer way to imagine creating a consensus around a meaningful shared vision. Figure 4.4 suggests the idealised scope of such an arrangement. The industry's clients and service providers are key, but need an axis with government policy leadership for the necessary motivation and action to occur. It is common to say the industry is unduly fragmented, but it is common too for

Table 4.1 Background information to the selected reports.

Organisation	Title	Reference	Country
nCRISP	Building future scenarios	Edkins (2000)	UK
	Constructing the future	Broyd (2001)	UK
	Nanotechnology and implications for products and processes	Gann (2003)	UK
CIRIA	UK construction 2010 – future trends and issues briefing paper	Simmonds & Clark (1999)	UK
CABE	The professionals' choice – the future of the built environment professionals	RIBA (2003)	UK
CII	Vision 2020	CII (1999)	USA
CERF	The future of the design and construction industry (projection to 2015)	Building Futures Council (2000)	USA
Chalmers University	Vision 2020	Flanagan et al. (2000)	Sweden
CRC	Construction 2020: a vision for Australia's property and construction industry	Hampson & Brandon (2004)	Australia

government interests in construction, such as planning, housing, construction itself, etc., to be located in different ministries and to have diverse remits and concerns. So, it is reasonable to suggest that the leadership provided by governments at a policy level often needs to be harnessed collectively so that a clear dialogue can be held with the other actors. The third axis is to education and research, primarily through universities. A wealth of relevant knowledge is available about practices worldwide, robust conceptual models, experience in other sectors and educational solutions for long-term change. This dimension completes a balanced set of groups that represent different perspectives and that can through a debate, not devoid of tensions, fashion a robust, shared vision.

Creating and maintaining a national strategy

This is precisely the process that is taking place, for example, in The Netherlands within their current major industry review (PSIBouw, 2004) and through the CCIC (2004) in Canada. A number of vision documents related to construction have already been produced in different countries over the last few years, and the most prominent of these have been analysed by Salford's Centre for Research and Innovation in the Built and Human Environment (SCRI, 2005) and are summarised in Table 4.1.

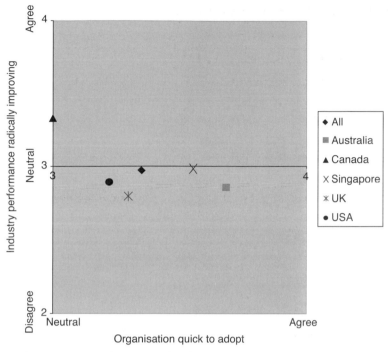

Figure 4.5 Organisations' adoption and industry's improvement.

VTT (2005) reviewed 16 construction industry strategies (some sectoral), and found a strong consensus on drivers and a wide range of key issues, but that 'implementation is barely covered' (p. 10). This links with the findings of the questionnaire survey in five countries (Barrett & Lee, 2004) illustrated in Figure 4.5, namely that, although construction organisations appear to be quick to adopt new initiatives, this has not led to radical improvements of industry performance. The whole would appear to be *less* than the sum of the parts. A reactive industry is not automatically an improving industry if the energy available is not spent in synergetic ways, hence the importance of a vision.

Creating a consensus vision between key stakeholders is hard enough, but maintaining this over time and achieving progressive implementation is harder still. As Ang (Ang, G.K.I., 2004a) points out, the process of reform will take longer than the life-time of one Parliament. In this report, Ang illustrates the fragility of political support with experience from: Australia where the Construction Industry Development Agency had only a three-year life (by design) from 1992–95; Denmark where Project Hus, a very major initiative, was terminated after only two years on a change of government; and Singapore where participants in the PSIBouw consultation felt the Construction 21 initiative had been losing momentum since its inception in 1999 and was becoming 'routine'. The overriding impression from national reform experience in several countries is that it is hard to start, but then a range of initiatives can be implemented. However, after around three or four years the political context weakens or the initial energy begins to fade, at which point the process can easily collapse unless it is reinvigorated in some way. The experience in the UK is that maintenance of the national reform push has been through a number of cumulative initiatives, e.g. Latham (1994), Egan (1998) and Fairclough (2002). It is symptomatic that CRISP was re-named nCRISP in 2002 to indicate 'newness', and was succeeded in 2005 by the industry's High Level Group of its Construction Technology Panel. This links to an analogous group at a European level. Ang argues that probably a ten-year concerted push at a national level is a reasonable expectation of the effort required to achieve significant change in the industry.

This dynamic is illustrated by the procession of initiatives in Norway, recounted by Tom Rellsve of RIF (PSIBouw, 2004): experience of partnering through the oil and gas industries in the early 1990s; a study of the construction sector in 1993; and a programme of research from 1996 to 2001 leading to a 'basis for thinking', but little take-up. Then in 2001 a major series of workshops that identified practical ways forward and achieved buy-in from a range of stakeholders has since led to a range of practical ongoing developments. This is a positive story. However, at some point it seems reasonable to hope that the changes will become embedded in the industry, and maybe this is typified by the emphasis in Australia shifting after 2000 to engendering industry leadership of 'learning networks' designed to make the changes self-sustaining.

This area is analysed in more detail in Chapter 11, around the Dutch experience of trying to balance vision with political reality, against a background of a lack of trust and fundamentally conflicting ideas about the roles of competition and collaboration. This national process was informed by an audit of international experiences and highlights the complex dynamics of these processes.

Achieving sustainable impact

It is interesting to stand back from these processes and consider appropriate models for the major, society-level change processes being discussed. It has become quite common to quote Gladwell's (2000) 'tipping point' model in discussion about how to move from a strong consensus amongst a few to a more general, epidemic, uptake by the many. The sub-title 'How little things can make a big difference' and the notion of different people with different skills taking part in the process is attractive, if those 'little things' can be identified and those different people can be found and mobilised. This report and the strategies summarised above are aimed at identifying the few high leverage factors that can nudge the current equilibrium into a new position and so underpin major improvements in construction. But who will carry the message?

Gladwell argues for connectors with extensive social networks who spread the idea, informed by mavens (experts) with an obsessive knowledge about the topic in question, and reinforced by salesmen with a flair for persuading others about the idea. Of course various individuals can exhibit more than one of these characteristics, but the message for construction change is that once the vision is bubbling up amongst the key stakeholders an imperative for implementation is to put together a complementary team that can ignite interest amongst those throughout the industry. The Australian Cooperative Research Centre for Construction Innovation (Hampson & Brandon, 2004) is an example of a concerted effort to engage large numbers in the industry through

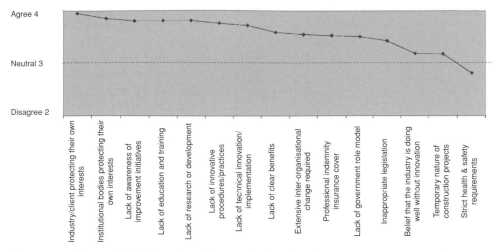

Figure 4.6 Factors inhibiting the widespread adoption of new construction practices.

a big programme of interactive workshops. Between them the core team behind this programme (John McCarthy the CRC chair, Keith Hampson its chief executive and Peter Brandon consultant to the CRC) strongly reflect a mix of the three 'tipping point' roles. Given the need to make sense of and respect the specific context of the many industry players (as argued in terms of sub-industries above) it seems likely that many teams will in fact be needed. As Gladwell puts it: 'that is the paradox of the epidemic: that in order to create one contagious movement, you often have to create many small movements first' (p. 192).

This contains within it an argument for working back out through the existing representative professional institutions that have typically been 'consolidated' to create a focus for the creation of the vision. These institutions once fully engaged have tremendous networks and membership loyalties on which to build. However, if they are not included and converted to advocates then it is timely to highlight that in the results of the survey of five countries (see Appendix 4), institutional bodies 'protecting their own interests' was the second highest rated inhibitor of the adoption of new practices (see Figure 4.6). The highest was clients 'protecting their own interests', and so it is also key that a client voice is actively encouraged and engaged with.

If Ang is right that ten years' effort can be expected, then it is reasonable to think that such a programme will go through phases, during which different emphases can be detected that progressively build. Kotter (1996) proposes a generic step-by-step process for achieving implementation of change at the organisational level. As an experiment, the main steps of this model were reinterpreted and infused with 'interesting practices' from case study countries that appeared to apply to the Dutch construction situation (PSIBouw, 2004) as shown in italics in Figure 4.7. On the face of it the model provides a robust framework to accommodate the various initial actions that are in train, whilst also projecting forward into successive phases over the longer term.

This appears quite straightforward, but could at best represent only a rough guide to how a national programme of construction reform should be designed. Van de Ven

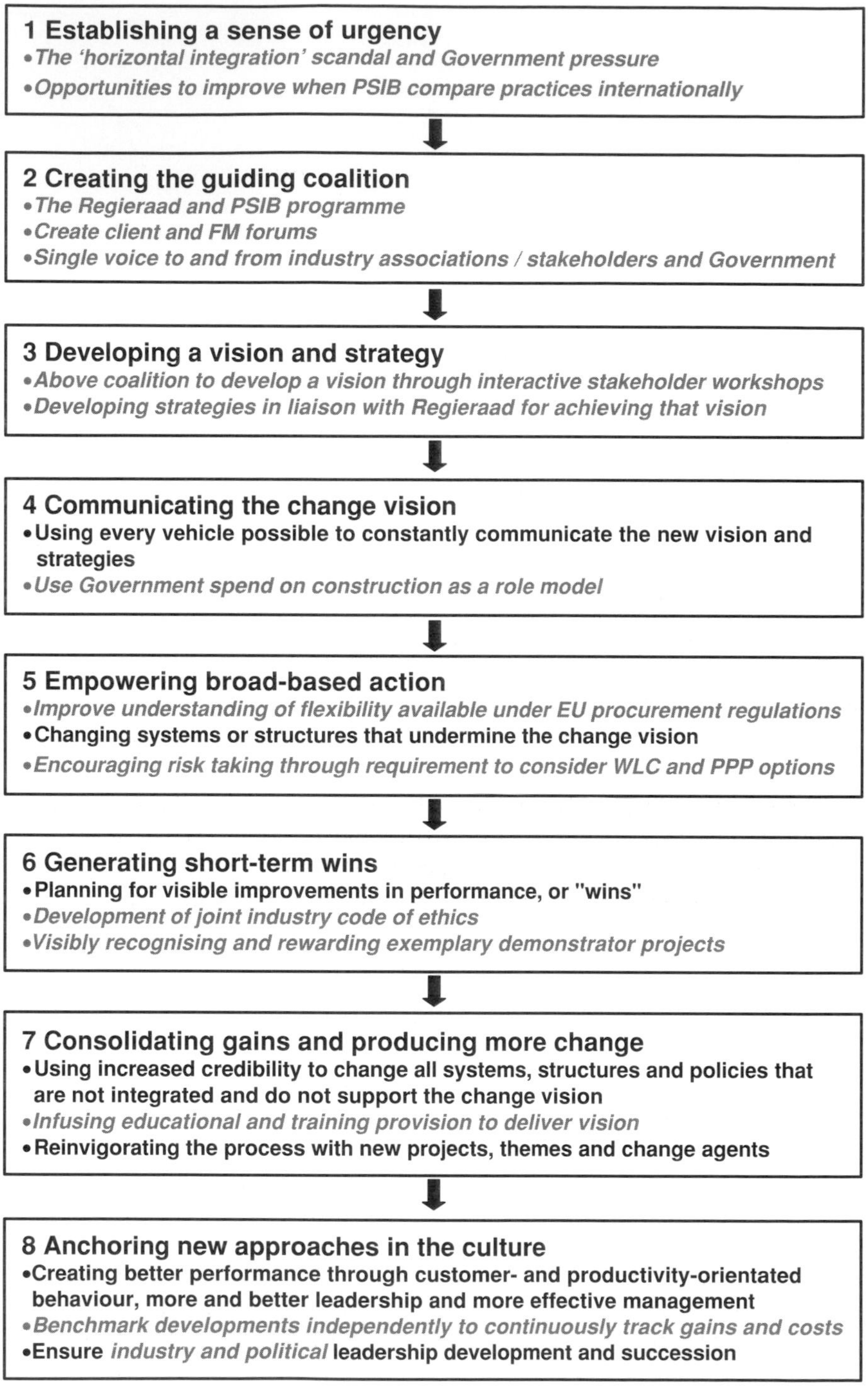

1 Establishing a sense of urgency
- *The 'horizontal integration' scandal and Government pressure*
- *Opportunities to improve when PSIB compare practices internationally*

2 Creating the guiding coalition
- *The Regieraad and PSIB programme*
- *Create client and FM forums*
- *Single voice to and from industry associations / stakeholders and Government*

3 Developing a vision and strategy
- *Above coalition to develop a vision through interactive stakeholder workshops*
- *Developing strategies in liaison with Regieraad for achieving that vision*

4 Communicating the change vision
- Using every vehicle possible to constantly communicate the new vision and strategies
- *Use Government spend on construction as a role model*

5 Empowering broad-based action
- *Improve understanding of flexibility available under EU procurement regulations*
- Changing systems or structures that undermine the change vision
- *Encouraging risk taking through requirement to consider WLC and PPP options*

6 Generating short-term wins
- Planning for visible improvements in performance, or "wins"
- *Development of joint industry code of ethics*
- *Visibly recognising and rewarding exemplary demonstrator projects*

7 Consolidating gains and producing more change
- Using increased credibility to change all systems, structures and policies that are not integrated and do not support the change vision
- *Infusing educational and training provision to deliver vision*
- Reinvigorating the process with new projects, themes and change agents

8 Anchoring new approaches in the culture
- Creating better performance through customer- and productivity-orientated behaviour, more and better leadership and more effective management
- *Benchmark developments independently to continuously track gains and costs*
- Ensure *industry and political* leadership development and succession

Figure 4.7 Kotter's change model applied to the Dutch PSIB initiative.

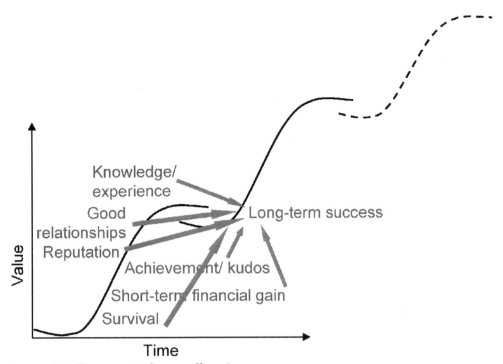

Figure 4.8　Firms must take as well as give.

et al.'s (1999) careful longitudinal study of long-term innovation efforts stresses that: 'the innovation journey... [is]... messy and complex... [and is typified] by a nonlinear cycle of divergent and convergent behaviours that may repeat itself over time' (pp. 212–213). They also stress that it is a process that can be manoeuvred but not controlled, and that the built-in tension between the diverse interests needed to achieve learning can probably best be resolved through pluralistic, distributed leadership (pp. 116–124). This is a good summary of this section so far.

To be sustainable the vision not only has to address maximising the joint value created, but also the equitable distribution of rewards to those involved. As Hennes de Ridder puts it 'money for value' in place of 'value for money' (de Ridder, 2001 quoted in Ang, 2004a). But the rewards are not only financial and immediate. Of course those involved in construction have to make a living, survive and ideally make a profit so that they can re-invest for the future. However, many appear to be operating at the absolute limit so that survival alone is the principal driver. To break out of this self-confirming cycle requires the mobilisation of additional returns to business that are non-financial, at least initially (see Figure 4.8). Those involved in projects can (and do) get back a sense of achievement and kudos associated with successfully completing projects against demanding requirements. They can also build a stock of goodwill, good relationships and high reputation amongst those they do business with. They have to learn strongly between and not just within projects so that the firm's knowledge base grows. Taken altogether, these softer benefits can be taken from current activity and fed

Figure 4.9   Creating value for all.

into the company's longer term success. The following sections will focus more on these aspects, but the main point here is the notion that without encouraging and valuing these additional dimensions the industry will tend to remain in an unrewarding rut.

Summary

In summary, the broad conception of construction set out in this section can inform the creation of a *shared vision amongst key stakeholders* for maximising value across the whole life-cycle of constructed artefacts. The act of building consensus visions for construction is evident to varying degrees around the world, and is a non-trivial activity in its own right. There is much to be learnt from experiences in Australia, The Netherlands, Singapore, the UK, etc. Creating initial momentum and keeping things moving is a complex, multi-dimensional activity and necessitates the engagement of a balanced range of interests. Together they can navigate the specific terrain in their country and develop a formula around 'creating value for all', in the sense that the value created is maximised for all stakeholders and that those who contribute are rewarded for their input (Figure 4.9). Taking this forward over ten or more years will necessitate revitalising the approach at various stages and engaging sub-sectors in more focused implementation activities. However, an over-arching vision that is robust, but regularly refreshed, is a key element to maintaining cumulative progress over the medium to long term. Such a vision will then become a sound basis for many other concerted actions within the industry, by government in its regulatory and client role and by the educational sector.

   Senge et al. (2004) stress how important it is to understand the relation between parts and wholes, and that the whole is realised through the parts. This move to a deeper, more general understanding of how individuals contribute to the broader effort is vital to making progress. The 'infinity' model, given earlier, linking seven key areas is intended as a key contribution on this front. It is hoped that it will enable people engaged with creating visions to take a robust, connected overview. It could be said that it is a seed from which it is hoped a mighty tree will grow. For this, as Senge and his colleagues note, more is needed beyond the key role of the seed:

It is common to say that trees come from seeds. But how could a tiny seed create a huge tree? Seeds do not contain the resources needed to grow a tree. These must come from the medium or environment within which the tree grows. But the seed does provide something that is crucial: a place where the whole of the tree starts to form. As resources such as water and nutrients are drawn in, the seed organises the process that generates growth. In a sense, the seed is a gateway through which the future possibility of the living tree emerges (p. 7).

Consensus visions are crucial, but continued engagement with those affected is essential too as engagement grows and the process goes through distinct phases, whilst also dynamically adapting to circumstances as they emerge.

5 Balance of markets and social capital

Peter Barrett

Introduction

Cut-throat competition, unscrupulous low bidders, lack of trust and poor risk management are very commonly quoted characteristics of construction. These are implicitly
reinforced by standard contracts, custom and practice and the sheer confrontational
nature of the industry. Do the benefits accrued justify the direct and opportunity costs
involved? Does it have to be like this? The core issue is the pre-eminence, especially
in the public sector, of the lowest cost basis for awarding construction contracts. From
this flows the fragmentation of the participants involved, both between and within
projects. Findings from the questionnaire survey of five countries (see Appendix 4)
reaffirm that issues such as existing procurement routes, public policies and the structure of the industry have a very significant influence on the value gained by parties,
but they do not strongly support its equitable distribution (see Figure 5.1).

Some theories

The generic question about the 'pay-offs' associated with the pursuit of short-term
personal gain is addressed in the 'prisoner's dilemma' experiments with their roots in
game theory (Kay, 1993, pp. 35–49). These are metaphors for making choices within
business relationships. The basic scenario is that two prisoners are arrested and placed
in separate cells. Although the prosecutors have no real evidence, the prisoners are
given the following options. If one confesses he will go free and the other can expect a
ten-year gaol sentence; if both confess, each will get seven years; if neither confess then
both will probably get, say, a one-year sentence on a trumped up charge. The 'best'
choice for the individual depends on the other prisoner's choice. The logical decision
for each prisoner in this situation is in fact to confess so that he or she has a chance
of getting off completely or at worst serving a seven-year (not a ten-year) sentence. Of
course what is missed is the opportunity to cooperate and both agree not to confess so
losing the chance of getting off completely, but securing light (one-year) sentences for
both prisoners. The dimension of cooperation has been explored through an iterated
version of the game in which the 'prisoners' have past experience of the other's actions
and future expectations of continued interaction. This in itself does not change the
calculation, but experimental evidence shows that people do mostly reach cooperative

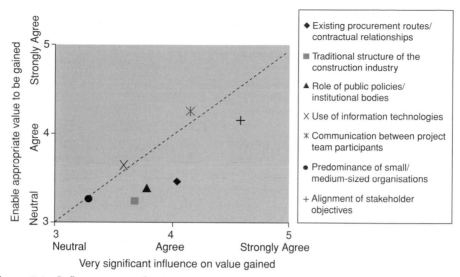

Figure 5.1   Influences on value in construction.

solutions. The main features of the successful cooperative strategies in iterated games are that the participants: begin by expecting the other player to cooperate, not that he or she will cheat; and they respond to bad behaviour and punish it, but not too severely; and they are forgiving.

This has echoes of McGregor's (1957) Theory X and Theory Y, in which it is clearly argued that if you start with negative assumptions you will inevitably create the behaviour you fear. Conversely, if you start the relationship positively, it may not work out, but at least it has a chance and can be steered back on track if an aberration is dealt with proportionately. Of course a positive commitment has to be carefully chosen, but Kay highlights the power that such commitments can have using the illustration of William the Conqueror literally burning his boats on landing on English soil in 1066. This reduced his options, but also those of the English – William was not simply going to go away, he had to be taken seriously! Kay goes on to argue that the general lessons from game theory are that rather 'old fashioned' aspects of business, such as reputations, commitment and relationships, are not optional self-imposed constraints, but rather they represent mechanisms that evolved over centuries to deal with problems in social and commercial relationships.

Interestingly at the broader societal level, Sacks (2003, pp. 149–160) makes a parallel argument for valuing relationships borne of ongoing interactions. These, he argues, will be created at a local level within groups of people that share a bond that goes beyond individual transactions and could be called 'social capital' or the level of trust in a society. Sacks takes the argument beyond the benefits for contractual dealings and distinguishes these from what he terms 'covenantal' relationships. These are often evident in family life, but in broader life are sustained by loyalty, responsibility, fairness, compassion – professionalism at its best. Without these trust-based, reciprocal relationships and the institutions that support them, 'markets and states begin to fray . . . social life itself loses grace and civility'. Particularly relevant in the public sector, Sacks argues that: 'excessive centralisation inhibits the growth of civil associations, just as excessive commercialization erodes them. The proper balance is precarious and hard to maintain'.

Releasing value in relationships

So, how does this relate to construction? The commercial environment for construction has been typified as 'competition is good; more competition is better' (Ang, 2004b). This resonates with Hendriks' phrase: 'the poison is the dose', that is, something that in moderation can be beneficial, in too great a quantity can be fatal (Hendriks, 2004). Given these aggressive conditions it is hardly surprising that 'lock in' is felt by the individual players. Who can afford to move first? Well maybe the answer is the clients, but why should they? Some interesting work has been carried out by Zaghloul & Hartman (2003) in which the effects of 'disclaimer clauses' in contracts between clients and contractors were studied. These clauses are extensively used in traditional and new partnering style contracts to shift risks to the contractor on issues such as delay and uncertainty of work conditions. The study was based on a survey of more than 300 industrialists in North America. It revealed that contractors assess the five most commonly used of these clauses such that a premium of between 8% and 20% is added to their price in a seller's market. This is a very clear and tangible measure of the significant cost of risk within construction relationships. Interestingly, the authors found that low levels of trust were typical, but that where a high level of trust did exist the impact on the premiums was profound. The premiums in these circumstances were 'very low' because the perception of the shift in risk was from around 4.4 to only 2.2 (on a five-point scale). The benefits of a high level of trust were found to be facilitated if the parties had previous experience of working together, such that they felt the other party displayed competence, integrity and, more generally, a good reputation within the industry. In ideal, but rare circumstances the risks were intelligently and equitably distributed without recourse to blanket clauses. So, for clients the value of building trust is clear, and the route forward through some form of ongoing relationship has been highlighted as a significant practical variable.

Some movement away from the domination of the lowest price regime towards the creation of longer term relationships within the contractual domain has been played out in many ways and continues to evolve. The phrase 'partnering' or, in Nordic countries, '*samspill*' (playing together), typifies many of these approaches that seek to create strong teams with broad stakeholder representation from the start of the project. A study of practice in various countries for the Dutch Government (PSIBouw, 2004) highlighted the prevalence of 'partnering' approaches, but also that they were commonly underpinned by traditional contracts. Not exactly indicating commitment by burning any boats! The label of partnering can in fact be a cover for a 'new' approach that is not always significantly different in its level of trust or commitment beyond the base contractual position. A parallel development has been the increasing use of public–private partnerships to finance the construction and operation of public sector buildings, typically over a 30-year life span. Here the commitment from the start is orientated towards the use phase, and the consortia involved are compelled to address this. On a broader front, in Norway at least, government legislation is being used to make it mandatory to consider life-cycle costing in the design of new public buildings. A similar requirement to consider 'partnering' has been legislated for in Denmark since January 2004.

The consensus from the workshops in five countries was that the key objective was to procure flexibly, optimally and appropriately. This is a complicated way of saying that just using the 'normal' approach, because one always does, is not good enough. Procurers need to assess each situation and be flexible enough to establish the optimal approach that is appropriate for the given situation. This will be conditioned by many things, but is likely to take into account the value of long-term relationships. This can be realised through pre-qualification using a range of key performance indicators (KPIs), which can include the track-record or capacity of the players to work well as a team. Selection based on broader criteria than simply price is becoming quite prevalent, albeit in two-stage tendering price is likely to dominate in the second stage. However, in one Norwegian project, for example, a psychologist's assessment of the bidding consortia's make-up as a well-functioning team was decisive, on the grounds that a good team would be best able to dissolve the problems found. Linked to these selection issues, there is also the need to set up the project processes so that they are performance orientated and reward behaviour that is aligned to project needs. Taken together these types of action can place the project participants in a very different and much more positive context than is normally found, but they can also resemble 'paper armour' if done superficially.

Ignoring the various labels being used, some of these approaches are not so novel or risky as they may appear. For example, it is worth remembering that 25 years ago in the UK, for example, construction professionals had fixed-fee scales and competed on quality and reputation. This seemed to work without too many problems before Government pressure via the Office of Fair Trading swept this approach aside in the name of free markets. Incidentally, Ofori (2004a) highlights a converse lag in some developing countries where old UK contracts are now being institutionalised, when in fact the newer approaches would be much more in tune with the local cultures. On a more general note, the oligopolistic market for smaller projects in most towns is such that the same few people are interacting in ongoing relationships that go well beyond the specific projects, extending to joint engagement in general professional and social networks. Good behaviour springing from these covenantal relationships can often be seen, but has rarely been studied (Swan et al., 2004). This social capital should be valued and other extra-contractual aspects, such as a strong professional and ethical stance, should be encouraged. For example, accountants in Hong Kong have clear ethical guidance (HKSA, 1997) and, more specific to construction, the main professional associations in Norway have established a joint code of ethics that covers a balanced range of issues as given in Table 5.1.

The role and influence of professional associations has been reduced in some ways by the dominant role of the markets. However, this analysis indicates a very positive role for them, provided they can escape their protectionist inclinations and work together. It was noticeable in the workshops in five countries that it was the contractors who were most exercised to argue for the importance of professionalism. Public

Table 5.1   Topics covered by the Norwegian Code of Ethics.

Legality	Fair competition
Environment	Cooperation and mutual respect (*samspill*)
Profit for all	Balanced contracts
Satisfying client's requirements	Conflicts of interest avoided
Justice and respect for employees	Discrimination avoided

sector clients in particular seem to feel hemmed in by probity demands and, for those working in Europe, EU regulations are perceived as restrictive. For all that, good examples of appropriate and seemingly optimal approaches are to be found. In the private sector it seems clear that the pedantic use of lowest cost tendering is avoided, but of course there will be times when it is appropriate, say for simple projects where a number of competent and trustworthy suppliers are available. To underpin progress in practice in this area those procuring at an operational level need to be strongly supported. Appropriate contracts for the different situations need to be available with clear guidance: norms about what is acceptable need to be broadened through clear messages from influential sources and high-profile examples. Finally, multi-dimensional benchmarking frameworks could be beneficial, within which information about the performance of the players in construction can be gathered. This could be along the lines of the monolithic, single model being developed by the Building Evaluation Centre in Copenhagen. There are alternatives that are looser, but could be just as effective.

For example, feedback and ratings are collected and made available for hotels around the world via open web-based facilities such as 'Trip Advisor'. Is it impossible to imagine such an approach working for construction? Trends in ICT capabilities and usage move quickly and users readily develop the capacity to assess the ragged range of views made available. A similar development could be expected on web-based auctions along the lines of e-Bay. And indeed this approach has already been used by a major international bank for a building contract, saving £0.3m in the final few minutes (see Figure 5.2). Of course there are complex implications running from such innovations, some quite problematic (RICS, 2004), but moving *with* social trends is not a bad rationale for going forward and at least there should be experimentation with these approaches.

Within the whole area of procurement is the key issue of risk allocation and management (CII, 2004). This is particularly clear in studies of mega-projects, which highlight the importance of embedding risk and accountability within the decision-making processes, especially at the Go/No-go point of the project (Flyvbjerg et al., 2003). This attention to risk management should also be evident as each link in the 'supply chain' is forged, through careful partner selection that includes consideration of important risk areas, such as health and safety (HSE, 1998). Active risk management must then run through the rest of the project, ideally using 'cooperation' models with the objective: 'to increase the ability to take risk – and not only the ability to avoid risk' (Artto et al., 2000).

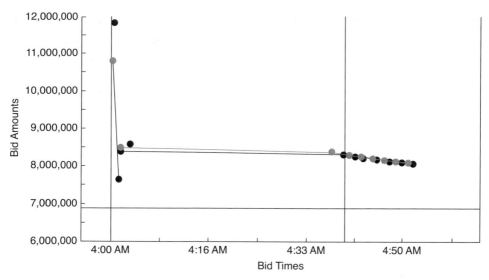

Figure 5.2   Electronic reverse auction.

Summary

So, in summary, there is a major question surrounding roles and relationships and the extent to which the balance of markets and social capital is optimal at present. The predominance of market thinking seems to lead to a 'lock in' situation where it does not appear logical or even possible for any one actor to change, even though it is pretty evident that under current arrangements all suffer. Even taking a strict market view this can be sub-optimal, as calculations of the cost of *not* having trust illustrate. The role of longer term perspectives and relationships is vital here. Partnering/alliancing arrangements are a move in this direction, building on some good existing work (Scharle, 2002; Consultants, 2004; Liu et al., 2004; Swan et al., 2004). But a move to less contractual, more fundamental shared social norms, rooted in professionalism and codes of ethics, is an area that deserves more attention. Practical outcomes are likely to be the selection of project participants based on all-round performance, not simply lowest price, and an equitable allocation and management of risk. For this to happen accessible and appropriate norms, metrics and contracts will be needed.

It should be remembered that the above prescription is based on a proposed movement from the confrontational *status quo*. However, it is quite possible to move too far in the direction of cooperation, as the recent problems of collusion, or 'horizontal integration', in The Netherlands (PSIBouw, 2004) have illustrated. This issue has also been picked up in the generic management literature in terms of the 'dark side of close relationships' (Anderson & Jap, 2005). Thus, the emphasis on 'balance' in the title of this chapter.

Chapter 12 explores this notion of balance further through the use of 'scenarios' and also picks up on the far-reaching impacts on procurement of a shift to a service orientation, already pinpointed from an economic viewpoint in Chapter 3.

6 Dynamic decisions and information

Peter Barrett

Introduction

In tandem with the above market considerations, there is a need for dynamic decisions and information throughout the building life-cycle. Currently there are huge gaps in the process through which information and understanding are lost at very great cost. Decision-making is undermined and this is compounded by the combative culture of the industry.

Drawing on the revaluing construction workshops in five countries, Figure 6.1 indicates the dependence of actions in the project domain on positive procurement behaviour and shared information and regulatory frameworks across the industry. Without a conducive context those working at the project level are almost certain to adopt defensive routines that minimise their risk, but close down opportunities to maximise the joint value created. However, once these conditions have been realised, in whole or in part, then significant progress can be sought through increased integration using ICTs and the adoption of an appropriate team value management approach. The common theme to these two foci is seamless integration, focused on, respectively, information/technology and people/decisions.

Information and technology

In the area of information and technology a report by NIST (Gallaher et al., 2004) dramatically highlights the potential for improvement by assessing the costs of inadequate interoperability in the US capital facilities industry. The headline figure revealed is $15.8bn 'lost' every year, or 1–2% of industry revenue. The report covers only commercial, institutional and industrial facilities, but the analysis extends for the whole life-cycle, from planning and design, through construction to operations, maintenance and decommissioning. A clear estimating methodology is given based on interviews and surveys. Interoperability is defined as 'the ability to manage and communicate electronic product and project data between collaborating firms and within individual companies' design, construction, maintenance and business process systems'. The estimates were arrived at by comparing the costs of existing practices with 'a hypothetical counterfactual scenario in which electronic data exchange, management and access are fluid and seamless'.

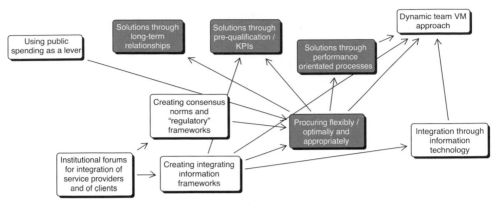

Figure 6.1 Interactive actions in the 'project' arena.

The results indicate that two-thirds of the 'unnecessary' costs are met by owners and operators of buildings, whereas architects and engineers are only associated with about one-tenth of this impact. The balance of around one-quarter of the costs is met by the industry in terms of contractors, specialists and suppliers. The clear implication raised in the report is that: 'interoperability costs during the O+M phase [result] as a failure to manage activities upstream in the design and construction process'. The figures of course also indicate that motivation to move on the issue is skewed, in that the greatest potential to change things is in the area where the least cost is experienced. However, interviewees expressed the view that, although a failure to exploit emerging technologies was an issue, it was a series of major 'disconnects' in the processes that was the major disincentive.

There have been various proposals for removing the unhelpful disconnects in the construction process. Lahdenpera (1995) sets out an ambitious 'holistic' analysis, whereas Kieran & Timberlake (2004) argue in a more focused way for manufacturing methodologies to be taken up to achieve a position of 'fabricating' buildings rather than constructing them. Lee et al. (2000) have been working collaboratively with UK construction stakeholders to take generic product design concepts and apply them to the construction industry. The outcome has been the 'Process Protocol', which takes a whole building life-cycle view and endeavours to identify hard and soft gates, or progress points, together with a continuous archive-building activity within which experience is collected. This framework has helped some collaborators to take a fresh look at their activities and optimise some of the transitions. Davidson & Dimitrijevic (2004) highlight the need for more 'information about information' in construction and distinguish generic from project-specific information. They suggest three types of information, namely: people-related, technological/physical and product/process. For each of these the challenge is how to keep a flow of general and project information interacting at the right points and flowing cumulatively throughout the project process and building cycle.

The results of the questionnaire survey in five countries (see Appendix 4) particularly highlighted the significant potential for positive interactions between project

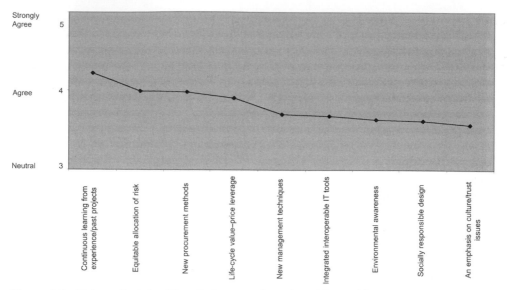

Figure 6.2   Drivers that significantly increase the value delivered by construction.

experience and organisational learning to lead to increases in value delivered by construction, as shown in Figure 6.2. Interoperable IT tools are also shown in this figure, but alongside other issues such as risk management and procurement. As mentioned above, these latter aspects, as well as constituting potential drivers for progress can, if they have not been addressed, hold back the initiatives being discussed in this chapter. So, the picture seems to be one of clear routes forward being identified, but requiring action on many fronts to be achieved. *Assuming* the construction market issues covered in the previous chapter have been addressed, the question still arises as to how feasible is the counsel of perfection implicit in the initial NIST analysis?

First, there is the issue of the availability of appropriate ICTs and the skills to use them. In general, Cairncross (2001) makes clear that there are rapidly rising global trends of internet use. Further, it will 'increasingly be conventional businesses that find imaginative ways' (p. 102) to use IT, rather than specialist 'dotcom' companies. There is also a trend towards a common language with three-quarters of mail in English. More people have English as a second language than as a first (p. 280). It is apparent from everyday experience in the developed world that the use of mobile phones and the internet are rapidly becoming commonplace for everyone, young and old, rich and poor. In the developing world there is a stark contrast, but also the opportunity to skip technologies – there is evidence that this is happening. So for many the technology and skills are in place at a generic level, but can they be beneficially employed in construction?

Aouad has been working for many years with colleagues and industry collaborators to develop nD CAD systems using virtual environmental interfaces. The nD refers to the move from three-dimensional tools, through 4D with the time dimension added, to multi-dimensional systems that can make available information about space and time for construction, and also, for example, costs, legal issues and operational factors. The system developed has been trialed in the waste-water industry and in the speculative housing sector (Gallicon, 2001). It has successfully satisfied stakeholders that it can deliver real benefits in terms of what-if modelling early on, demonstrations to potential customers, acceleration of the overall process, better value management, cost savings and, crucially, the capacity to build libraries of objects from one project to the next. This technology has since been taken forward and has proved itself in real-world applications. Another industry-collaborative project, focused this time on reactive, minor maintenance works, which makes up around 30% of UK construction. This highlighted both the tremendous waste in the, amazingly complex, current information processing systems and the clear potential, confirmed by the industry partners, to eliminate much of this waste (Barrett et al., 2003). For example, a typical transaction cost to raise a payment was estimated at £50, but the average size of the job was only £100; however, the clients had thousands of such jobs each year. The web-based knowledge management system created greatly reduced the inefficiencies of the existing systems and opened up the possibility of 'a learning system' that could become more and more powerful over time.

So it does appear that the general capability for using ICTs can be built on with effective systems for construction. However, so far these systems have been driven by well-motivated, major clients. For the systems to come into general use will be more of a challenge, but work on common standards and a move towards making systems available through 'free-ware' should make progress on a broader front increasingly feasible. There remains the question of content though and the key strands that should run through the various systems.

Turning back briefly to Chapter 3, on a holistic idea of construction, the notion of the 'use phase' as the ultimate focus of all other activities was raised. Preiser & Vischer's (2005) book on assessing building performance provides a good benchmark of thinking in this area. The concept of linked building performance evaluations being made at all stages of the building life-cycle is promoted (p. 17), with feed-forward implying that the knowledge so generated should be captured by the organisations involved for more general use. Szigeti et al. (2005) highlight the notion of 'fit' between what is wanted by clients and how it is provided by the industry. Another concept promoted is that of 'universal design', that is: designing for the great majority of the population and their variable capabilities (pp. 170–178). In addition, the book gives information in the form of case studies and advice on the various roles involved and the many practical tools available for different forms of evaluation at the different stages. A loosely related concept is that of 'open building', in which the stress is on the ongoing flexibility and adaptation of facilities to meet user needs through a never-ending design process. This area concerns CIB Working Commission 104 'Open Building Implementation' and there are extensive reports of case study projects using this type of design thinking. As noted before in, for example Norway, legislation has been passed that requires whole-life costing issues to be taken into account for buildings funded by the public sector (PSIBouw, 2004).

The CIB has also been putting significant emphasis behind the notion of 'performance based building', that is: 'the practice of thinking and working in terms of ends' (Gibson, 1982). A special journal issue that focused on this topic provides a benchmark of the current position (Szigeti & Davis, 2005). There is a strong emphasis on regulatory change and the technical performance of buildings. There has been significant change in various countries on these fronts; however, the practical implementation of this approach and the actual costs and benefits are surprising. Take up has been very slow (Bakens et al., 2005), and simply placing the opportunity for innovative designs in front of the industry can lead to bad mistakes being made in some instances (Duncan, 2005). The lessons highlighted are that as well as technical/regulatory progress, significant effort has to go into training and engagement of those in industry and of clients, and that until this has been achieved the scope in high-risk areas should probably be constrained to some extent.

Szigeti & Davis (2002) clearly feel that the time is ripe to make the link between a growing body of information about what makes buildings function well for users and the briefing process. However, they are also concerned that 'the barriers in the system will work against the results of the Probe studies, and other similar efforts, being accepted and adopted in the mainstream by all stakeholders' (p. 51). Jaunzens et al. (2001) see the facilities manager as a key agent for the introduction of use-phase orientated data at all stages of construction. They define the stages using both the traditional RIBA Plan of Work and the Salford Process Protocol (pp. 25–27). For them the ideal is that the building is not just delivered as an artefact, but as a working facility. However, they acknowledge that for this to be achieved facilities managers need to have developed 'comprehensive benchmarks... reviewed and interpreted in the context of corporate strategy' (p. 23). This links with the notion of design quality indicators, debated in a special journal issue by Dewulf & van Meel (2004), Eley (2004) and Slaughter (2004) amongst others. Issues of hard 'scientific' or soft perceptual measurement are considered, together with questions of professional versus layperson's involvement. The prominence given to the costs and issues of ongoing maintenance and use vary. This is clearly an area for further debate and development.

The above use-phase orientated information and dimensions should be considered and if possible facilitated by the nD ICT systems being developed. In particular, it would be excellent if multiple stakeholders could use such systems, not just as black boxes that calculate a 'correct' answer, but as enabling environments for better joint decision-making. This observation links well with the analysis by Hansen et al. (1999). They studied knowledge management in management consultancies and distinguished between two principal strategies, namely 'codification' or 'personalization'. The aim with the codification approach is to 'Provide high quality, reliable, and fast information-systems implementation by reusing codified knowledge'. 'Personalization' aims to 'Provide creative, analytically rigorous advice on high-level strategic problems by channeling individual expertise' (p. 109). 'Codification' assumes a volume of work with much reuse of knowledge by large teams consisting of many juniors, using people-to-document systems, highly supported by IT. 'Personalization' assumes high margin work carried out by well-qualified small teams using person-to-person knowledge management, supported by moderate IT systems. But which alternative relates best to construction? To choose a predominant strategy (codification or personalisation, respectively) depends on whether: the service is standardised or customised; the organisation/sector mature or innovative; and the knowledge used to solve problems explicit or tacit. Much of construction is customised, company systems are of relatively low maturity (Construct IT, 1998b) and the knowledge used is predominantly tacit.

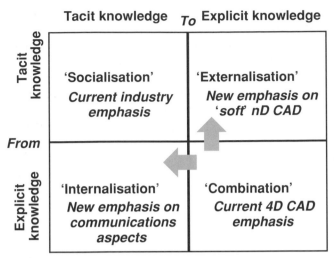

Figure 6.3   Modes of knowledge conversion.

This all points towards a strategy that emphasises teams, person-to-person knowledge management and only moderate IT support with an emphasis on communications. It is not possible to generalise, but it seems that construction faces the uncertainty and, doubtless as a consequence, has many of the characteristics of a new, dynamic, thrusting industry. Construction may be old, but it is not 'mature'! This links in an interesting way with Tenner's (1996) diagnosis that pushing systems further and further in terms of sophistication and detail will inevitably lead to problems, and that the solution is 'finesse'. So, this means aiming for systems that cover the important hard and soft data dimensions and use technology and ingenuity to make the resulting system accessible, easy to use, robust, integrative, dynamic and flexible.

Underlying much of this discussion has been the distinction between hard and soft data, which parallels to a degree the distinction between explicit and tacit information. Figure 6.3 (based on Nonaka & Takeuchi, 1995) uses this classification and endeavours to summarise the optimum emphasis of nD CAD. If the *tacit–tacit* emphasis of the industry is taken as a given, then the mis-match with the *explicit–explicit* character of 4D CAD systems is stark. For the industry to operate to its optimum the synergies of all four modes of knowledge conversion are needed (Barrett & Sexton, 1999). The implication is that 4D CAD systems need to shift emphasis towards the *tacit–explicit* mode by accommodating a wider than normal range of key hard and soft, long- and short-term performance criteria (nD CAD) at a coarse, but robust level of resolution. In parallel with this, a push towards supporting *explicit–tacit* knowledge conversion is needed with an emphasis on richer communications.

The developments suggested should create a closer fit between systems' characteristics and the reality experienced by those in the industry. As such it will simply make more sense for such systems/approaches to be taken up through industry pull, especially if an incremental, but progressive trajectory is supported (Barrett, 2003). The sense of technology not being a major driver for change, but more needing to reflect the characteristics of the industry was evident in the first phase of consultation for the CIB Revaluing Construction initiative. As Courtney & Winch (2003) summed it up: 'the issues around construction performance improvement were not technological but organisational and behavioural' (p. 178). This same feeling came through the workshops in five countries carried out subsequently.

The discussion will now move from an overall view of the building cycle, emphasising information and technology, to a consideration of the specific construction stages of briefing and construction with a focus more on people and decisions.

People and decisions

Briefing ('programing' in the USA) is an area that has long been debated. The reason is clear, in that briefing is (Barrett & Stanley, 1999): 'the process running throughout the construction project by which means the client's requirements are progressively captured and translated into effect' (p. 134). As such, it is the crucial dialogue that underpins the construction process. It is most intense towards the start of the project and so decisions made have a huge impact later, but it should also continue iteratively, but progressively, throughout the project so that the resulting building is continuously fashioned to meet the client's requirements. The question of those requirements and how they are arrived at has been the cause of much debate between those who want them fixed early so that the remainder of the project is on a firm foundation and those who argue for a more interactive, extended process before even the main parameters are set. An example of the former stance is encapsulated in the RIBA Plan of Work (1967), which foregrounds the inefficiency of iterating after certain points. In contrast, Othman et al. (2004) reinforce the value of a continuing process if the final product is to match or exceed the client's expectations. They argue for a 'dynamic brief development' approach and identify a set of drivers that support this notion. They also interestingly point out that if the 'traditional' approach is taken and the brief artificially 'fixed' early on, then all that happens is that the confrontational 'change order' regime takes over during the latter stages to provide the flexibility needed.

Horgen et al. (1999) argue that briefing is concerned with 'work practice' as it is situated within four interdependent dimensions of the workplace, namely: space; organisation; finance; and technology. Their case studies track a highly non-linear process that moves forward, but cycles between various levels and events, including a slice that focuses on designing the process of briefing that is being undertaken itself. They describe a rich, team-based approach that highlights various challenges, including keeping existing organisational processes 'unfrozen' long enough to allow the optimal built solution to emerge. Luck (2002) has worked on a close analysis of the language used during interactions with users in briefing, and concludes that useful knowledge can be revealed. She highlights that the tacit dimension especially depends on a social process of discussion where metaphors and stories are valuable vehicles. She concludes that: 'design knowledge cannot be completely represented in a prepositional, non-contextual form' (p. 16). Barrett & Stanley (1999) argue for the necessity of allowing time for trust to develop so there is time for co-learning through linked processes of disclosure and feedback. So, briefing has to be a dialogue. How extended and open-ended this is will depend on the novelty of what is intended, and also on the knowledge, experience and degree of mutual understanding of the main participants. It is of course essential that the above is not used as an excuse for procrastination. Nutt (1988) suggests maintaining a strategic stance to the process, but stresses that: 'any attempt at a wholly rational process must be considered to be both impossible and also unwise' (p. 135) given the pragmatic nature of the situation confronted with limited information and the imperative to act. His approach highlights important aspects to be given attention, such as the distinction between factors that are reversible/irreversible, controllable/uncontrollable and predictable/unpredictable.

In summary, briefing is clearly a key process, but to confine it to only the early stages and as a 'factual' position statement is a mistake. Briefing is an ongoing, social process through which joint understanding is developed, political and financial constraints are negotiated and creative solutions are hopefully found. It also has to be pragmatic, progressive and broadly cumulative, retaining a strategic stance whilst dealing with a wide range of issues, big and small. So defined, the briefing process articulates strongly with the building production phase, which is the next area for discussion.

Rethinking or re-engineering the construction industry as a production system has been a lively topic over recent years. For example, stimulated by radical improvements in manufacturing industries, the International Group on Lean Construction has been active in testing whether the concepts and techniques can be beneficially applied in construction. Luis Alarcón's (1997) edited book of papers established a benchmark after three years' work by the group. Ballard & Howell (2003) have been at the forefront of promoting 'lean construction', and at its simplest who could argue with 'maximising value and minimising waste' (p. 119)? They stress that some characteristics of a lean construction approach as opposed to a non-lean process are: a focus on all product life-cycle stages; the involvement of downstream players in upstream decisions; and not performing activities until the 'last responsible moment' (p. 122). In principle these are attractive propositions consonant with much of the briefing discussion above; however, they do cut across a lot of traditional professional and contractual boundaries and may have their limits in terms of practical or appropriate application.

Lean and other new approaches were the subject of a special journal edition (Winch et al., 2003) and there is a mix of views. Koskela (2003) argues that more than structural change is needed, that a new theory of construction is needed and that this should emphasise more the operations level. Ballard & Howell (2003) suggest the superiority of 'lean' production principles, whereas Winch (2003a) queries their appropriateness to the construction industry. Barlow et al. (2003) argue that mass customisation as in Japan may be a paradigm for house building, but not for other market segments, which is a useful reminder of the notion of sub-parts of the construction sector with distinctive characteristics. Gibb & Isack (2003) focus on pre-assembly, although they show that clients do *not* confirm that they perceive benefits from this approach. Arbulu et al. (2003) provide a case study of value stream analysis, within which it is revealed that 96% of the time spent was non-value adding; however, they acknowledge that re-engineering in construction has had mixed results. Green & May (2003) raise an important query in their argument, that by focusing so strongly on delivery to customer requirements and only that, there is a danger that over time this undermines the capacity of the industry to do just that! As they put it: 'the relentless focus on technical efficiency blinds industry leaders and researchers to the damaging side effects. The casualization of the construction workforce raises real questions about the industry's long-term capacity to deliver high quality construction' (p. 104). So, there are many opportunities, but a danger that an unduly scientific, rarified, dehumanised approach could be taken at the extreme. This is a consistent theme with the briefing discussion where the process can be in danger of overwhelming the people and relationships involved.

At a less radical level the benefits of value management (VM) were often mentioned at the workshops in five countries. Judging from an international practitioners' workshop (CPN, 2004) VM is widely used in various private and public sector organisations. The approach has been defined as: 'a structured approach that supports the decision-making process by addressing the key success factors and criteria from a functional standpoint' (p. 1). In a survey of VM use in Hong Kong (Fong, 1999), again it was found to be in quite wide use in the public sector, but its spread was being constrained by a lack of trained personnel. However, those who had first-hand experience were very positive, and it was felt to be essential for capital projects to both reduce costs and improve quality. A VM industry workshop (SCRI, 2005) illustrated the link to 'best value' in the public sector. Through Kelly's 'three wheels of best value', general quality improvement initiatives in the local authority are linked to VM of specific projects mediated by a panel of experts who assess the services being delivered and whether any new services are required (Kelly & Hunter, 2003). This linkage between organisational and project processes is interesting, in that it extends beyond the project perspective that is very much the stress of the lean, etc., approaches.

However, it is necessary for VM to impact during the whole of the construction process and this is reinforced by Neff (1998) in his advocacy of the STEPS approach, where the usual political, economic, social and technological factors are supplemented by a demand to find 'synergies'. Neff's argument is illustrated by a case study of a complex sub-service project, and gives a recommendation that the STEPS factors need to be reviewed and risks managed progressively with clear mitigation measures and controls. It is in these latter areas that the potential for synergies becomes evident, especially as the practical complexities and constraints mount through the project period. In general, little has been said about continuity beyond initial VM efforts fairly early in the project, although many from the design side of the industry see VM workshops as horrendous, destructive events. The challenge must be for VM to become a more natural part of the thinking and practices of those involved, rather than a confrontational, occasional event. Neff stresses that for this type of joint problem-solving to work, goodwill and trust are demanded. It is timely to refer to Doz's (1996) findings about collaborative endeavours, that is: alliances develop through cycles and against the three criteria of efficiency, equity and flexibility. Doz makes the telling point that 'the impact of initial conditions quickly fade away' in successful alliances. This again argues against the rigid definition of a project at an early stage and for building the capacity, skills and trust that allow problems to be jointly resolved against broad performance specifications. For small, straightforward projects it is quite possible that the VM activities will be really quite informal. However, the notion, highlighted in the briefing discussion, of regularly returning to refresh the shared understanding of the client's aspirations, and steering the project against these, still makes good sense.

Summary

In summary, large amounts of resources are wasted through the disjointed processes used in construction. Many of the activities appear not to be value-adding. General ICT trends mean that the opportunities to address these problems are available, including the capture of learning by organisations from project to project. Within projects joint, dynamic, decision-making using value management techniques has great potential at varying levels of formality. This can lead to new ways of collaborating and thus the realisation of the full creative potential of those involved. Taking the relevant products of these processes right through to the use phase is important, as is infusing the processes with a 'building performance' emphasis. It is important to add that these logical propositions all have a lot of potential to contribute to realising maximum value through construction, but at each point they need to be tested to ensure that they appropriately address the human dimension: tacit knowledge; informal processes; iteration to allow re-assessment. Similarly, longer term impacts beyond the project must be considered so that the capacity to manage for value is not lost, but rather is enhanced over time.

Chapter 13 takes on the above issues in the context of nD (multi-dimensional) modelling, which appears to be a key way in which technology can enable significantly improved information flows and decision-making over the whole building life-cycle.

7 Evolving knowledge and attitudes

Peter Barrett

Introduction

The areas for development set out in the last two chapters could have a big impact on particular projects. However, for sustained and widespread change in the industry it is important that there is a general evolution in the knowledge and attitudes of those involved. The need for specific skills has been mentioned, but also the capacity of 'thinking in wholes' must be enhanced. The workshops in five countries uncovered profound concerns about the workforce available for construction, for example: a trade skill and, looming, age gap; a lack of design and construction management skills; and a lack of top management capacity. In terms of clients, there were particular concerns about the industry's failure to understand them and their lack of ability to clearly express their requirements to the industry. Of course there are good clients and good practitioners, but the general level is of concern. In terms of universities, it was felt that they were not generally targeted at industry needs. The implications of this analysis are highlighted in Figure 7.1.

Educating top management at a level commensurate with their responsibilities is a priority. Further, continuing training and development at all levels are essential, focused around the skills needed to adopt improved practices and to take a broader view. The box concerning 'educating about clients' covers both ends of the relationship. Clients need educational opportunities to be better clients and a client-orientation should infuse the development of all construction personnel. To support the linkage between universities and industry there should be joint forum/s so that a dialogue can be maintained and a mutual understanding of the participants' distinctive roles allowed to develop.

Educational context

There are very few powerful drivers behind the educational aspect of the picture. However, action here is critical to the success of the overall revaluing construction effort. People are the industry's most valuable raw material. Dainty et al. (2004) go as far as to call the situation in the UK 'the construction labour market crisis', that demands collective action. Knowledgeable clients have a key role to play, as does government, as both can use their influence to reinforce the involvement of educational providers in key forums.

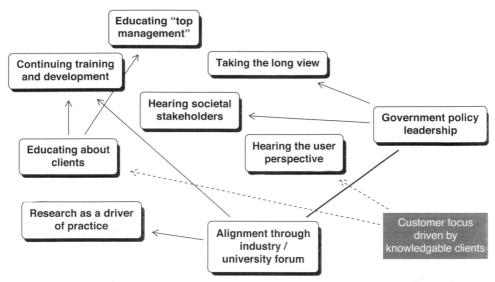

Figure 7.1   Dependent actions in the 'knowledge and attitude' arena.

In Chapter 14 Lees provides an interesting overview of construction education in the context of revaluing construction, but taking a long-term perspective. This sets out something of the shape of the construction-related education sector and its characteristics and suggests some areas for development. On a qualitative front he points out that construction education generally came late to universities and, as a result, it would seem it has made strenuous efforts to prove its academic worth. There have been two consequences: first, the syllabus is packed with respectable subjects, such as law and economics; and, second, the learning process reflects the engineering emphasis on lectures. So the courses explicitly claim to educate, leaving practical training mainly to the industry. This raises significant issues that are explored further in Chapter 14.

In terms of numbers of workers in construction and the demand for new trained personnel, Lees highlights that there are around 111 million construction workers worldwide, even on the limited SIC definition (see Chapter 3). Lees indicates evidence of strong growth projections for construction activity worldwide. This must translate into demand for new recruits in the high-income countries given the already high levels of productivity achieved. In low-income countries it is likely there will be a qualitative shift towards a more highly qualified workforce that can exploit more advanced technologies. In either case, high-level training is likely to be key to the industry responding effectively to demand. Unfortunately, in the UK at least, there was a halving in the number of construction graduates from 1995/96 to 2002/03. This reflects a decline in the popularity of construction-related careers amongst those going into higher education. Having said that, from 2002 to 2004 there was a sudden up-turn in interest, reflected in around a 33% increase in first-year registrations. This could be due to targeted advertising by the CITB, but other factors are likely to be at play too as construction is competing with multiple alternatives. Even so, it is easy to see how a shortage of trained personnel has come about and, given the lags in the cycles of education, it will take

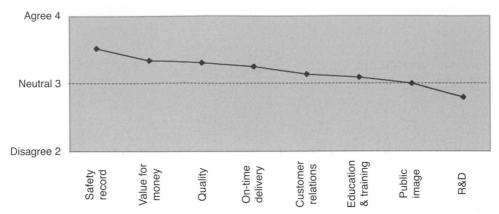

Figure 7.2   Where is construction radically improving its performance?

many years to turn around. This increases the importance of the continuing training and development of staff, such that the existing workforce is enhanced and the lowest price mentality, that industry accuses clients of, no longer extends to their employees. Instead, striving for competitive advantage through a smarter, more loyal workforce should become commonplace.

Knowledge and innovation

This would require a positive orientation towards knowledge management and acquisition in the industry. The importance of active knowledge management is stressed by many (e.g. Carrillo et al., 2003). However, judging from the results of the five country questionnaire (see Appendix 4) research and development is *not* an area of improvement, especially compared with more immediate (and arguably important) demands, such as safety (Figure 7.2). Absorbing knowledge often happens unseen and unmanaged, as 'distributed knowledge' accumulates amongst the individuals in a firm (Larsen, 2001). At a company level, the adoption of new routines and processes in construction tends to happen through a gradual, low-risk migration (Barrett & Stanley, 1999) that is both pragmatic and also probably very sensible, given the systemic complexity of construction. Simply taking up 'best practice' is impractical and, as Gratton & Ghoshal (2005) highlight, unlikely to lead to achieving high performance. For this they argue company-specific 'signature processes' are needed that build on the history and distinctiveness of the company.

The CIB brings together a worldwide community of built environment researchers. They have a wealth of knowledge about global practices and trends and insights into other sectors. How can this activity best interact with those in practice? Houvinen (2004) argues strongly that an emphasis on 'business-management' research is needed to link theoretical findings to the application domain, and generic notions to the construction situation. Pietroforte & Constantino (2004) tracked changes in the papers in two leading US construction journals and noted a shift towards including company issues as well

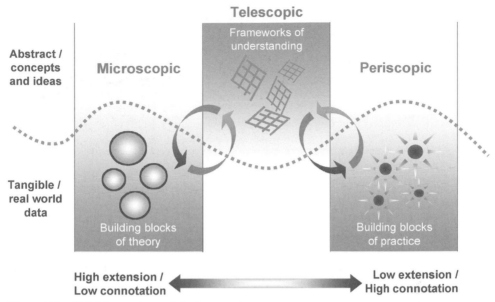

Figure 7.3 Kaleidoscopic model of research.

as traditional engineering topics. However, there is some way to go according to the workshops in five countries where research is not identified as a major factor for driving revaluing construction, but has a role to play in revealing inspiring possibilities and providing evidence of what appears to work. In this way innovation in processes, materials and business models can be supported and the transfer of good practice accelerated. The question of how academics can best work with industry is complex. There is only about one academic for every 10 000 practitioners in the UK, so there are very practical limits and it is important that the maximum value is derived from the interaction, which means those participating should make distinctive contributions. Barrett & Barrett (2003) set out a framework in which academics provide a reflective complement to the practical knowledge of practitioners. The framework highlights three approaches to research, as shown in Figure 7.3.

'Periscopic research' is where building blocks of practice are created through interaction with industry, studying what they are doing in the real world and endeavouring to make sense of it. This links to 'telescopic research', where an effort is made to create broader frameworks of understanding within which the building blocks of practice can be drawn together. This aspect is more theoretical, but as well as informing practice it is tested by it for relevance. The 'infinity' model of revaluing construction is a contribution of this sort.

The question remains of how best should practitioners work with academia to take away value from the activity. A study by Barrett & Gilkinson (2004) followed the impacts of collaboration in research by a snowballing series of interviews starting with those directly involved. It was found that beyond the information that could be read in a report, the first-hand interaction meant that they gained a much richer understanding of the issues. A number went back to their roles in industry with sufficient confidence to be advocates for change in their own companies, and in several cases set up as consultants to 'sell' the ideas to others. They also had the confidence to adapt the ideas to make them appropriate to their particular circumstances. Thus the prolonged interaction as collaborators had created individuals who could and did act as vectors for knowledge transfer. There are echoes here of the 'tipping point' approach described earlier. Davey et al. (2004) reported on the use of action learning with small construction companies and the contribution to industry change that this enabled. The core of this exercise was action learning 'sets' of practitioners who provided mutual support and stimulus within a structured environment. In both these cases it can be seen that a multiplier effect on the researchers' input is achieved through a social process. Given the numerical problem highlighted above these seem to be promising avenues to explore.

Summary

To summarise this chapter, for revaluing construction to be self-sustaining in the longer term an evolution of knowledge and attitudes is needed. There are few major drivers for this, and so governments and leading clients have to provide support to enable the close involvement of academia in strategic forums involving industry. The implicit message behind this in itself would raise the importance of objective knowledge to the whole process. Additionally, the opportunity to affect those individuals coming into the industry through their education is great, with economic growth in most countries leading to high demand. However, this provision must also link to continuing training for the bulk of, existing, employees. The nature of construction education could be brought much more in tune with industry needs if a challenging element of training was included, as in the high-status professions of medicine and architecture. All of this educational input should stress the importance of understanding clients and should extend to clients themselves. Lastly, research is a very rich way of companies co-learning with, the numerically few, academics. Thus, social processes that carry the learning more widely throughout the industry are needed. The comprehensive vision of revaluing construction, namely to maximise the joint creation of value and the equitable distribution of rewards, expanded into the seven action areas of the 'infinity' diagram, will only make progressive impact if those in the industry internalise the messages, and here the educational dimension is vital.

8 Awareness of systemic contribution

Peter Barrett

Introduction

The scope and size of construction is huge, as indicated in Chapter 3 on the 'holistic idea of construction'. What does *not* automatically run from this is an awareness of the diversity and systemic nature of the impacts that arise from this scale of activity. Economically, built wealth created through construction accounts for a major part of man-made capital. For example, in the UK this amounts to around 70% (around half of which is in housing) with the balance held in machinery, etc. (Pearce, 2003). The significant share of GDP has already been highlighted (but see Chapter 10 for more detail). There is then the question of the economic effect of construction on the rest of a country's economy. Chang & Chien-Chung (2004) studied developments in Taiwan from 1979 to 1999 and concluded that the data: 'clearly point to uni-directional causality running from construction activity to economic growth' (p. 591). This economic dimension has a human face, in that it is estimated that around 111 million people are employed in construction worldwide. This is a large group of people for whom construction provides incomes, livelihoods and careers. Many of these people work in small and medium-sized companies and two-thirds of them are in the developing world, even though only around one-third of construction output is from these regions (see Chapter 14 for more detail).

Contribution at a variety of levels

Of course construction results in physical artefacts and these are typically (but not always the case) created in a regulatory context that seeks to ensure minimum standards on aspects of public concern, such as structural stability, health and safety, energy conservation and accessibility. However, beyond these minimum standards there is a potential to create spaces that, to varying degrees, enhance the quality of life and competitiveness of those using the spaces created. So moving beyond the minimum standards of basic accommodation leads into considering the physical, functional and psychological effects of building spaces on those who occupy them. This is explored in detail in Chapter 15, but raises issues of how to capture measures of these effects, with post-occupancy evaluation as the traditional approach. However, Vischer and colleagues argue for a broadening of this notion to a 'building performance evaluation process model' that tracks the whole building life-cycle and includes reviews from the users' perspective at every stage including, of course, design. It can be difficult for measures of value such as this to be available beyond specific organisations and settings, although the facilities managers' community would seem to be an important, but probably under-utilised, vector for its transfer. But there are general studies that provide pointers. For example, in the area of higher education, CABE (2005) provide feedback from case studies of five UK universities and highlight the important design dimensions for the recruitment, retention and performance of staff and students. Interesting differences in views are revealed as staff appear to be more attracted by buildings with a 'feeling of space ... aesthetic appeal', whereas students stress more functional aspects: 'modern design ... quality of facilities' (pp. 46–47). A report by Eclipse (2002) sweeps in a range of such effects as 'intangibles' and they indicate in their literature review the pervasiveness of benefits to individuals and organisations, but also the difficulty of 'valuing' them. They suggest hedonic pricing, contingent valuations and choice experiments as possible ways forward.

So, going beyond the perceptions of occupiers and actually identifying complex, interactive causal links with physical and behavioural outcomes is difficult. This is exemplified in Heerwagen et al.'s (2004) paper on creating collaborative learning environments. This concludes that the central dilemma remains that: 'spaces designed to increase awareness and interaction also increase the potential for interruptions and distractions' (p. 525). However, a rare and good example of work that does cut a way through some of these issues is provided by Zeisel et al. (2003) in a paper that identifies environment–behaviour links between the design features of care facilities for patients with Alzheimer's disease in the USA and behavioural health outcomes, such as anxiety, withdrawal, depression and aggression. In general, these dimensions of behavioural health problems are seen to recede in the face of design features such as providing for privacy, variability in social spaces, careful (camouflaged) exit design and creating a residential rather than an 'institutional' feeling environment. But there are distinctions to be made: for example, verbal rather than physical aggression appears to correlate with the environmental design. This type of work provides very valuable insights and, although it was for a particularly vulnerable group, similar, less stark, issues would doubtless translate, say, to the office environment.

Overall, it can be seen that there are many counterbalanced, interactive effects, which reinforce the necessity of creative, elegant design solutions to achieve optimal built environments. Where homes, offices and factories can be designed to make their occupants happier, more efficient and effective, the positive knock-on effects on the economy in general are enormous. Where people can be made healthier and happier the broader benefits to quality of life in society can be substantial too. Here an open approach is needed that considers all aspects of the environments created to fully release huge latent synergies: for example, extending beyond simply colour to connect with aspects such as the visual arts and to music (Staricoff, 2005) and to the pervasive impacts of green spaces (CABE Space, 2004).

Going beyond the specific building, organisations can use their estates to reflect, and to an extent create, their distinctive culture and image. It is interesting to compare the estate strategies of two Boston universities, namely Harvard and MIT. In a fascinating record of the development of the Harvard estate (Nason et al., 1949), the comment is

Figure 8.1 Eiffel Tower, Paris.

made that 'education to survive must be given form and substance'. The concerted effort to create the 'collegiate way of living' led to the building of the 'houses', for self-contained communities of scholars that today make up Harvard Yard. This typifies the University, reflecting a powerful ethos and long tradition. Nearby MIT is another world-class university, and its estate is being developed to create and sustain a very different image, that of modernity and cutting-edge activity. The new MIT Stata Center is a clear manifestation of this, but it also embodies ideas such as flexible interactions via meandering streets and spaces, allowing accidental meetings. Organisations use buildings, but buildings reflect strongly on their organisations.

Widening the scope beyond specific organisations, built artefacts make major impacts on societies. For most people iconic structures such as The Eiffel Tower (Figure 8.1) immediately provide a sense of place and for many in the locality, one would imagine, pride. Soaring cathedrals and ancient remains have long represented expressions of people's history, sense of place and aspirations. This has resulted in attention on the

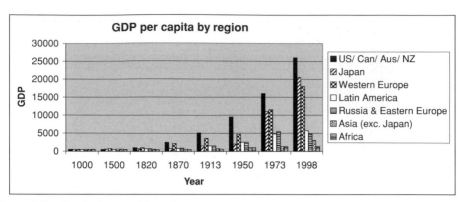

Figure 8.2   Gap between rich and poor.

built environment as a major way to regenerate depressed areas of nations. At their best such construction projects, as a tangible representation of commitment by government or other funders, give a sense of direction and involvement, lead to local jobs and long-term improvements in prosperity, raise the cultural vibrancy of the area and act as a catalyst for additional and ongoing private sector investment. An example of this in the UK is the regeneration of the old docks at Salford Quays where the Lowry Theatre acted as such a catalyst. Of course this notion of urban areas as a focus for analysis is crucially important and links to planning issues, which are central to making individual developments add up to attractive spaces/areas for everyone.

Some people are beginning to coin the phrase 'Urban FM' to highlight the need to actively manage some of these spaces too, to gain optimum impact. An extreme example of this is the Disney Corporation's experiment that started in 1996 with proposals to create a new town, called 'Celebration', for 20000 people. All available information on best practice/best value was used, including for the management structure. This is applied Urban FM, with 'a private company providing for the public good and under the control of customer orientated professional managers' (Roberts, 2004, pp. 28–29). This has been very successful in some ways, with house prices 20% higher than nearby communities and a waiting list to buy houses. However, there have been tensions, for example over control of amenities such as schools. Roberts identifies the central issue as one of how to represent the public interests of citizens when the consumer model breaks down. New vehicles such as public interest companies are emerging, but despite these difficulties the reason there is an issue at all is that there is clearly significant value locked up in the more general environment that is shared by individuals, and attention on recognising this more explicitly would be valuable.

The above example is a lifestyle community, and as such the town does not have many of the 'normal' problems of deprivation, etc. to deal with. This belies the huge long-term dynamics that underlie societal change. Figures 8.2–8.4, produced by the OECD and the UN, were kindly provided by Hamish McRae of *The Independent* newspaper in the UK and highlight some key trends. Issues such as the sharp divide between the rich nations and those that are considerably poorer may have opened over 1000 years, but the acceleration in the last 50 years has been phenomenal and raises

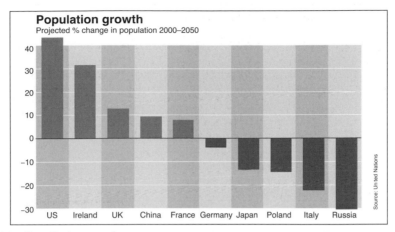

Figure 8.3 Population trends.

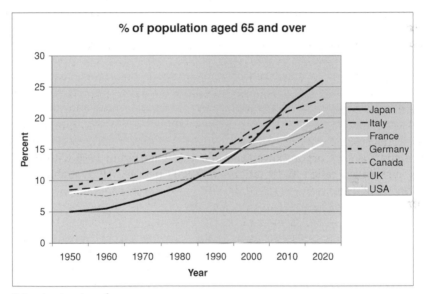

Figure 8.4 Ageing trends.

serious political and practical issues. Demographic trends are very varied, with population growth from 2000 to 2050 predicted to swing from over +40% in the USA to –30% in Russia. Within these figures there are major shifts in age profile with, in the developed world, an increase in the percentage of the population aged 65 and over from around 8% in 1950 to around 20% in 2020. These massive shifts in society act over decades, but of course the built environment, the output of construction, also exists in this sort of time-frame. The implications for the built environment are profound, from creating housing of at least the minimum standard for the world's poor, to adapting the built environment to huge quantitative and qualitative population changes. For

Figure 8.5   Building in Venice.

example, an ageing population implies different household sizes and adaptations to support the elderly living independently in their own homes. If this is not addressed there are major implications for their quality of life and a large bill for formal healthcare provision (Lansley et al., 2004).

Taking the still broader perspective of the planet and environmental issues, the built environment is responsible for: consuming 40% of all the raw materials taken out of the Earth's crust, by weight; 40% of waste streams; and 40% of greenhouse gas emissions (Milford, 2004). These are huge impacts and of course there are tremendous opportunities to reduce these impacts by more efficient processes, recycling materials and re-use. There are major moves to increase life-cycle and embodied energy performance, which should lead to reduced environmental impacts. The underlying rationale for these impacts is of course that construction leads to a harnessing of natural resources in such a way that habitable environments are created in which people can live and work.

Venice is a good example of a stunning city that could not exist without adapting nature to humankind's needs (Figure 8.5). The same goes for much of Holland, and sea defences around many coasts are key to societies living in safety. In areas afflicted by earthquakes, such as Japan, advanced technologies are used to create safe environments; and where space is limited, high rise, 'vertical' use of space is considered by many to be an elegant solution. So multiple benefits accrue through the way nature is adapted via construction, but there remains the necessity to optimise the cost at which this is achieved. Hart & Milstein (2003) suggest that companies can build a 'sustainability vision' by moving towards clean technologies, minimising waste and emissions and integrating stakeholders' views into their processes. They call this approach 'creating sustainable value' and emphasise the importance in this of shareholder value. For these authors this is a shift in thinking about sustainable development from: 'one-dimensional nuisance [to] multi-dimensional opportunity' (p. 56). The CIB itself has of course supported significant work in the area of sustainable development, resulting in agenda for both the developed and the developing worlds (CIB, 1999, 2002).

Summary

In summary, construction creates physical artefacts, and often the most prominent dimensions of this are the mess and disruption caused during construction and the cost of the works. Many of the benefits that derive from the efforts of construction occur over the longer term and thus are not automatically associated with the construction industry itself. There is a need for a greater awareness of the systemic contribution of construction. Highlighting the use phase links strongly to notions of construction as an investment and buildings as assets, in that users' physical, functional and psychological well-being are very dependent on the built environments they experience. This multiplies the impacts of construction through to every other aspect of the economy. Beyond the building perspective, construction is pivotal to urban regeneration and significant contributions are made to the cultural and environmental aspects of people's lives.

This pervasive stream of soft and hard benefits needs to be better accounted for so that the return on the initial investment can be made more clearly visible. In this connection key gaps that need attention are the creation of a comprehensive framework for these value streams and the clear identification of appropriate ownership amongst stakeholders. Chapter 15 explores this issue via the notion of dynamic 'building performance evaluation' throughout the life-cycle of the building.

9 Promotion of full value delivered by society

Peter Barrett

Introduction

The final area for action is the promotion of the full value delivered to society by construction. Of course this links strongly with the subject of the previous chapter. However, owing to the negative starting point for the industry, it is imperative that the contribution made is actively promoted to show that much of what construction does is good, exciting and creative.

The image of construction is under constant attack. For example, Figure 9.1 indicates the stream of UK governmental reviews of ever-increasing frequency that have been aimed at construction to try to work out how to fix it.

Evidence for strong industry performance

Maybe it is because of these reviews, or perhaps because the industry is in fact dynamic, that its relative gross value added has steadily moved ahead of manufacturing over the past 25 years, as shown in Figure 9.2.

Books, such as Woudhuysen & Abley's (2004) *Why is Construction so Backward?* jump on the critical bandwagon, albeit they are actually primarily critical of the planning system and have a vision of mass prefabricated housing and a new generation of skyscrapers. It could be of course that the industry is more in tune with what clients want than these authors. A survey by the NRC of major construction clients in Canada (Manseau, 2003) elicited responses from 205 companies, representing a 68% response rate. These companies are responsible for a large proportion of construction invest-ment in Canada and were asked to rate their satisfaction with their construction in-vestments over the previous three years. The results are given in Figure 9.3 and repre-sent a resoundingly positive view of the industry. For all that, it was the view from a high-level industry workshop in Quebec that grabbed the headlines, stating that con-struction in Canada was lagging ten years behind in terms of innovation (Cloutier, 2004).

A survey of 200 clients of building surveyors in the UK (Barrett, 2001) led to the results given in Figure 9.4. These indicate that clients generally assess the service di-mensions represented by the diamonds on the graph as midway between satisfactory and good. This is a robust performance, but with room for improvement, which should

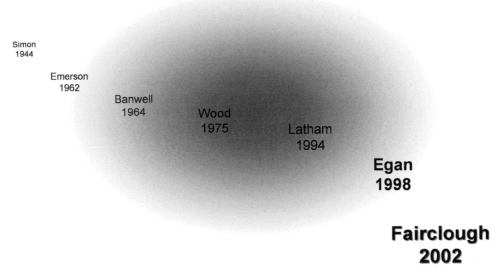

Figure 9.1   Stream of construction reviews in the UK.

clearly be targeted at the more important factors. So, Figure 9.5 maps the factors that are farthest below the diagonal line and highlights that understanding the client's problem is key, supported by timeliness and the ability to provide added value through service-orientated professionals. Thus technical correctness is taken as a given, but enhanced *service* is an area where improvement from a very solid base could be sought. Johnston (2004) stresses that a successful service orientation primarily involves being 'easy to do business with', the main aspect of which is dealing well with problems and queries. This is actually something that the strong project orientation of construction firms does typically deliver. However, a broadening of horizons is needed to resonate with the use-orientated conception of construction proposed by Carassus (2004), with its implications for the centrality of ongoing, interactive relationships between the industry and its client base. In Gronroos' (1994) terms there is a need for a paradigm shift from a 'transaction' model to a 'relationship' model. This is really quite radical for an industry that creates very tangible and enduring artefacts, but the point remains that it is how these products perform and adapt to productive use that is central.

This raises the issue of the dynamism of the industry. Using UK statistics over a 24-year period (Figure 9.6) it can be seen that employment in the industry has been highly variable. This reflects the level of demand in the economy and poses a major challenge for the industry, that is: how to cope with this level of turbulence. As Hillebrandt (2000, p. 27) states: 'the industry would do well to assume that it has to exist with fluctuations in demand and firms should take whatever defensive action is available to them'. The industry cannot buck the market, but how can it respond?

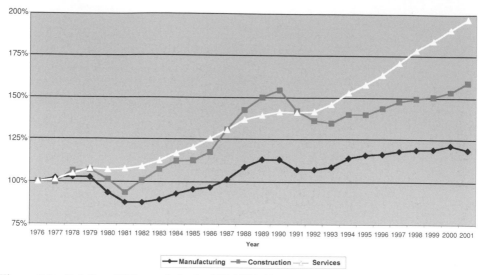

Figure 9.2 Relative GVA trends in the UK 1976–2001.

Senior Managers Satisfied With Construction Investments in
Last 3 Yrs?

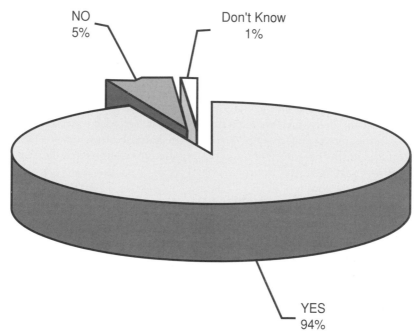

Figure 9.3 Satisfaction with construction in Canada.

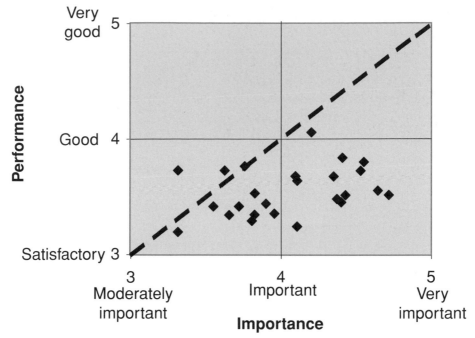

Figure 9.4 Importance of service dimensions versus satisfaction delivered.

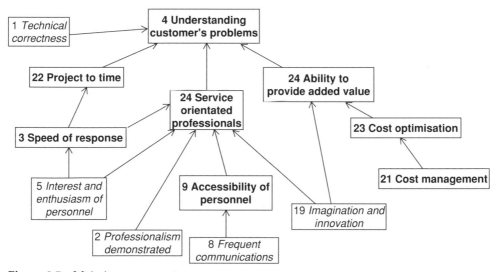

Figure 9.5 Main improvement areas.

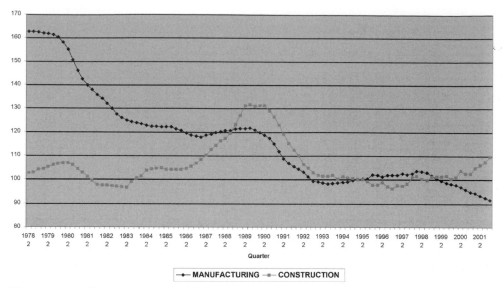

Figure 9.6   Relative employment in the UK 1978–2001.

It is no coincidence that the industry typically has very many small firms and a high degree of specialisation. This creates a high level of flexibility, but carries with it the need to be able to combine and recombine in different temporary groupings around specific project demands. The industry manages to do this well on projects ranging from changing a tap-washer to building an innovative multi-storey office block. A review of international practice (PSIBouw, 2004) gathered views from Australian policy leaders and there was a strong feeling that the industry had responded incredibly to the major increase in work that preceded the 2000 Olympics in Sydney.

It is strange then that construction is regularly portrayed as being less dynamic or innovative than other sectors. This is perhaps less surprising when the way construction is typified in economic industry classifications is examined more closely. Winch (2003b) makes the telling point that 'the bundling of "construction" goods and services used for the SIC is systematically different from that in all other sectors' (p. 652). He goes on to illustrate the point taking a generic value system running from design, through manufacture and distribute, to maintain. Then using the motor vehicle industry as a comparator it becomes clear that for this industry 'design' and 'manufacture' are included, but 'distribute' and 'maintain' are allocated to another sector. In contrast, construction SIC (F45) includes firms engaged in manufacturing, distribution and maintenance, but precludes design. So, architectural and engineering design services are incredibly classified out of the construction industry. This issue is discussed further in Chapter 10. However, the result of this is that the innovation profile of 'construction', as defined, is undermined by not including the inherently innovative design element and diluted by unusually including repair and maintenance work, which is less likely to be innovative, but in a mature economy accounts for around a half of the industry's output. Even in this latter part of the construction economy Sexton & Barrett (2003a) found evidence of a real capacity to move rapidly to exploit opportunities in small and medium-sized construction enterprises, albeit through mainly incremental approaches. Chapter 16 considers the concerns about the many small firms and the 'fragmentation' of the industry, but argues that this is in fact an asset, albeit with consequences.

The construction industry would benefit from joint promotional themes that draw on all the areas covered in the 'infinity' model. In doing this, the impression should come through strongly of a huge industry that is structured in a complex way to bring appropriate expertise to bear on a myriad of individual and societal needs. This means an industry where those involved feel a sense of pride and community, that dynamically develops holistic solutions for clients and users within reasonably stable frameworks, all underpinned by a strong professional and learning ethos and appropriate rewards. This would over time improve the environment, and in particular the starting assumptions, under which the sector labours. This in turn should improve the flow of high-quality people into the industry. The success of such a campaign would reinforce the richer conception of the industry, 'the Urban Space industry', as Jacques Rilling of CSTB in Paris has called it! An example of a campaign is the CITB's initiative in the UK. Its webpages entitled www.bconstructive.co.uk are orientated towards recruiting good quality people to the industry, and have been supplemented by an impressive television campaign that highlights innovative and iconic building projects. This television coverage has created a very positive image well beyond the target audience. The impact of this is described further in Chapter 14.

Summary

This section is the shortest of the seven areas for action considered. It depends to a great extent on the other areas becoming a reality, but it should also be grasped at an early stage to reinforce the significant 'good news' that the industry has to offer. As the other initiatives begin to bite they will reinforce and confirm the early messages.

It can be seen that a central part of appreciating the full value flowing to society from construction depends on taking a holistic perspective, thus strongly linking back to the starting point of this part of the book in Chapter 3.

Part 3
Views Around Priority Areas

10 Assessing the true value of construction and the built environment

Les Ruddock

Summary

This chapter builds on the first half of Chapter 3 and extends the treatment of the economic measures applied to construction. The traditional perception of the contribution of the construction industry to the economy is based on the methodologies employed for the definition and measurement of construction activity according to international standards. Within this context, the limitations of the concepts used in this definition are considered and an analysis is undertaken of the usefulness of the measures.

Construction activity has changed in response to new demands over recent decades and an evaluation is made of a new approach, to focus on construction activity to meet the changing needs of the economy and society. The role of built assets in the development of a nation needs to be considered. In addition, it may be that broader measures of the economic value of the built environment are needed to allow an assessment of the contribution of the built environment to quality of life, and to enable the value of the construction industry to be properly understood.

Introduction

Sustainable development has become an overarching ambition of national and supra-national governments across the world. It was enshrined as a goal of the European Union (EU) in the 1997 Treaty of Amsterdam, and subsequently shaped the commitment of the Council of Ministers to bring about economic, social and environmental renewal in the EU by 2010. The so-called 'Lisbon strategy' has been accompanied by various national strategies aimed, in essence, at improving the quality of life 'for everyone, now and for generations to come' (Office of Science and Technology, 1999). At the global level, the desire for sustainable development has been confirmed at various Earth Summits since the Rio de Janeiro Summit of 1992.

This desire has led to a fundamental review of our economic, social and other activities, and inspired the development of a number of policy initiatives. One of the key areas of concern has been the use and development of the built environment.

The built environment comprises buildings and infrastructure, and is one of the main components of a nation's wealth. Those structures and their collective placement play a major role in determining the quantity and quality of our activities, now and in the future. In most respects, they have had a positive effect on our quality of life, enabling the production of more and better goods and services and providing the basis for other creative social activities. However, they have also been a major factor in the production and persistence of numerous 'bads', notably environmental degradation.

Concerns about the use and development of the built environment, and the construction industry's ability to support sustainable development, have inevitably contributed to the notion of a review or revaluation of construction. The CIB Revaluing Construction initiative seeks to explore ways to improve the economic, social and environmental performance of the construction industry.

This contribution explores data on the gap between the actual and desired contribution of the construction industry and the built environment. In doing so, it raises a number of specific issues concerning:

- The definition and measurement of construction
- Problems of international comparisons
- The lack of data for broader analysis of economic value
- A framework for analysing the structure of the construction sector

Measurement of construction activity

The United Nations defines construction as comprising 'economic activity directed to the creation, renovation, repair or extension of fixed assets in the form of buildings, land improvements of an engineering nature, and other such engineering constructions as roads, bridges, dams and so forth' (United Nations, 2001). Construction activity represents a significant share of the economies of most countries in terms of its contribution to GDP and total employment, and it is also an important market for materials and products produced by other sectors of the economy.

A Standard Industrial Classification (SIC) is used to classify business establishments and other statistical units by the type of economic activity in which they are engaged. The UK SIC, is consistent with the general European and the international classification (ISIC) guidelines of the United Nations. All economic activity is allocated to specific divisions, groups and classes. The analysis of output by industry largely reflects the way in which the aggregate estimate is derived. The construction industry (Division 45) comprises: site preparation; building of complete constructions or parts thereof; civil engineering; building installation; building completion; and renting of construction or demolition equipment with operator. The UN guidelines distinguish between 'construction activity', which may be carried out by any unit irrespective of its predominant activity, and the 'construction industry', which is confined to those units whose predominant activity falls within Division 45 of the SIC.

This definition of the construction industry does not include other value-adding construction activities such as:

- **Upstream** – manufacturing; mining and quarrying; architectural and technical consultancy; business services
- **Parallel** – architectural and technical consultancy
- **Downstream** – real estate activities

Construction activity is conducted by a wide variety of agents, ranging from government departments and large private or public enterprises to small private construction firms and individuals. The typical structure of the construction industry in most countries is that of a highly differentiated industry in terms of a preponderance of small and medium-sized enterprises (SMEs). Using the developed countries of Europe as an example, SMEs represent over 90% of the total construction business in the EU and provide about 80% of jobs in the sector. In many countries, a significant proportion of construction activity is conducted by firms operating in the informal sector. The contribution of the narrowly defined construction industry to GDP ranges between 5% and 8% and between 5% and 9% of employment in EU countries (Euroconstruct, 2004). This, of course, understates the importance of construction activity, given that a significant proportion of construction activity is undertaken by units outside the construction industry.

The contribution of the construction industry to economic development

One aspect of the important contribution made by the construction sector to a country's economy concerns the relationship between that country's state of development and the level of activity in the construction sector. At the macroeconomic level, studies have tended to concentrate on developing countries (Turin, 1973; World Bank, 1984; Wells, 1987; Bon, 1990). The idea of an inverted 'U' relationship between construction activity and the level of income per capita (i.e. in the early stages of development, the share of construction increases but ultimately declines, in relative terms, in industrially advanced countries) was put forward by Bon (1992) and has been empirically tested by Ruddock (2000) and Ruddock & Lopes (2006). As noted by Tan (2002):

> In low income countries, construction output is low. As industrialisation proceeds, factories, offices, infrastructure and houses are required, and construction as a percentage of gross domestic product reaches a peak in middle income countries. It then tapers off as the infrastructure becomes more developed and housing shortages are less severe or are eliminated.

According to a study (referred to, in more detail, later in this chapter) undertaken by a CIB project group, after allowing for cyclical fluctuations, the general trend in construction activity in very developed countries is for construction activity to be in relative decline (Carassus, 2004).

In a study of over 70 countries categorised according to their levels of GDP per capita, cross-sectional analysis indicated that the share of total output rises but then diminishes with economic development. This is illustrated in Figure 10.1, where the construction industry's share of GDP per capita in 2000 is shown for four categories

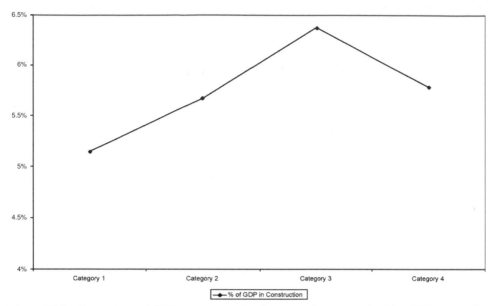

Figure 10.1   Percentage of GDP in construction for countries categorised by GDP per capita in 2000. (Categories: 1 and 2, less-developed countries; 3, newly industrialised countries; 4, advanced industrial countries.) Derived from United Nations (2003).

of country, grouped according to their state of development. Categories 1 and 2 taken together represent the less-developed countries, Category 3 the newly industrialised countries and Category 4 represents the advanced industrial countries. (More details of this study can be found in Ruddock & Lopes, 2006.)

Using data from the same study, a similar relationship can be found when the importance of the construction industry, in employment terms, is related to a country's level of income. The trend line shown in Figure 10.2 indicates that the industry begins to decline in relative importance in higher income countries.

A new approach

The case for a new approach to the valuation of construction activity has come from various viewpoints:

- The CIB *Revaluing Construction* focuses on improving the value of the final construction output and requires that the totality of activities involved in the production of the built environment is reviewed and, on that basis, recommendations are made for re-engineering or reorganising the process.
- The Preface to the Pearce report (2003) states that: '... the industry and its contribution to the UK economy and the health and well-being of UK society was neither fully understood nor adequately valued'.

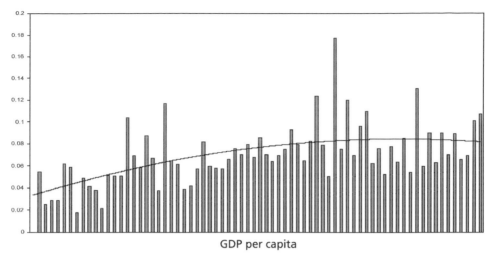

Figure 10.2 Proportion of total employment in construction in 2000 (for 71 countries ranked by GDP per capita). Derived from United Nations (2003).

- Carassus (2004) proposes a framework system approach for understanding the construction sector.

The rationale for this new approach is based on the view that the role of the construction sector should be viewed in a wider context than that of the narrowly defined ISIC definition of the industry. Figure 10.3 illustrates this approach and indicates the extent of the construction sector system.

The construction sector framework

In recent years the construction industry has come to play a new role within the economy of many developed countries, moving away from a production-based focus to one where it acts as a provider of services for the built environment. Carassus argues that large-scale production by the construction sector on housing and civil engineering projects formed a necessary part of economic growth policies in post-World War II industrially developed countries. Since the 1990s, however, there has been a notable change in terms of the demands placed upon the construction sector. Emphasis has moved away from building stock creation and is now placed on the repair and maintenance of the building stock created during the growth years. The increased importance of managing the existing stock is illustrated by the figures in Table 10.1. A research project was set up under the auspices of the CIB in nine developed countries[1] to test the application of Carassus's proposed framework. The results of the nine-country study showed that repair and maintenance work represented about one-half of the total

---

[1] The nine countries in the CIB 'Construction Industry: Comparative Analysis' study were: Australia, Canada, Denmark, France, Germany, Lithuania, Portugal, Sweden and the UK.

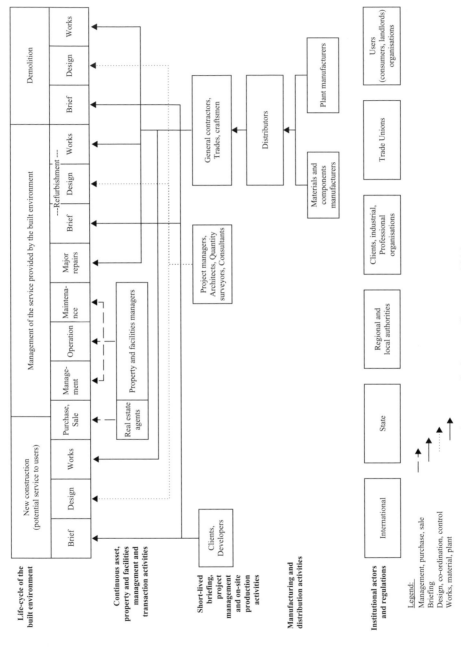

Figure 10.3 The construction sector framework. Source: Carassus, 2004.

Table 10.1 New construction and repair & maintenance (% of the value of total construction output).

	Australia (2002*)	Canada (1997)	Denmark (2000)	France (2002)	Germany (1999)	Lithuania (1999)	Portugal (2000)	Sweden (2000)	UK (2001)
New construction (Building)	36	41	33	41	51	20	68	25	44
Repair and maintenance (Building)	38	21	38	37	34	27	6	53	24
New construction (Civil engineering)	20	36	17	14	11	17	24	17	10
Repair and maintenance (Civil engineering)	6	2	12	8	4	36	2	8	23

(Source: Carassus, 2004)
Note: UK figures for building only include residential repair and maintenance, and the figures for civil engineering include all non-residential repair and maintenance.
* Estimated

73

value of the industry's work. The notable exceptions were Canada and, particularly, Portugal. (In the latter, this discrepancy from the general pattern can be put down to data availability issues.)

Beyond repair and maintenance, managing the existing stock has become a strategic issue for companies and governments. The impacts of facilities management and public–private partnerships are strong signs of this trend. Quality and maintenance cost aspects of the service rendered by the built capital have become important, as too has flexibility of use. This means that the operations and function of the construction sector need to be considered in a different manner. The industry has become much more involved in dealing with whole life-cycle issues, which have become the dominant feature of sustainable development.

The approach highlights the need to improve statistical data and to link the building and property sectors in economic studies. The main concern is data on the characteristics of the building stock and public/private in-house building and maintenance departments. This is critical for the role of the construction industry in improving the efficiency of this stock and its management systems.

The composition of construction activity: comparisons to find common trends

More evidence of the general trend towards a greater share of construction activity being in the repair and maintenance sector comes from a Euroconstruct study (2004). Euroconstruct[2] provide data on the construction industry in 19 European countries (Figure 10.4). A stock of over 170 million occupied housing units constitutes a permanent reservoir of work for the construction industry in the Euroconstruct countries. This has resulted in a dynamic demand in the housing sector as a consequence of various factors (low interest rates, increasing numbers of first-time house buyers, the growing trend towards secondary residences, tax measures in certain countries, etc.). Typically for European countries, the share of repair and maintenance work is now almost as large as the value of new construction work in these countries, constituting 47.5% of the countries' aggregate output in 2003.

In periods of recession and weak growth, repair and maintenance work plays a counter-cyclical role in the building industry, acting as a 'shock absorber' for the industry. The demand for repair and maintenance work depends on the extent of the existing building stock (which, for the last two decades has grown at an annual rate of approximately 1% in the developed economies of Europe) and on the number of national and local urban renewal programmes.

Figure 10.5, reflecting the situation in the 19 *Euroconstruct* countries, illustrates this aggregate effect, with the forecast virtual disappearance of growth in new building work in 2006 and 2007. This compensatory effect is even stronger in countries, where

---

[2] The 19 Euroconstruct countries are: Austria, Belgium, Czech Republic, Denmark, Finland, France, Germany, Hungary, Ireland, Italy, Netherlands, Norway, Poland, Portugal, Slovakia, Spain, Sweden, Switzerland and the UK.

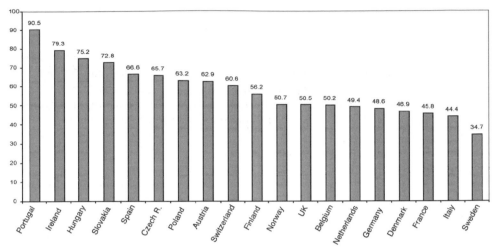

Figure 10.4 Share of new construction in total building (%) in Euroconstruct countries in 2003 (Mean = 52.5%). Source: Euroconstruct, 2004.

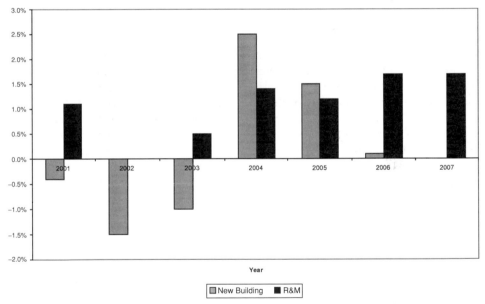

Figure 10.5 Share of new build and repair & maintenance (% change in value over previous year) [Actual, 2001–2004; forecast, 2005–2007]. Source: Euroconstruct, 2004.

the renovation of the existing stock now represents the majority portion of the market (The Netherlands, Germany, France and Italy) and those where it is on a par with new construction (Norway, UK and Belgium). The phenomenon does not come into play in countries where new construction predominates (Portugal, Ireland and Spain in particular).

The contribution of the built environment

The economic value of the built environment – traditional measure

The production of new buildings and other structures in any given period adds to a nation's economic wealth, in the form of the built environment – it also contributes to social capital. The nature of built capital formation therefore plays a major role in determining both the quality of life, and also the nature of sustainable development, both as a goal and as an instrument of government policy. A focus on improving the value of the built environment requires that we review or analyse the totality of activities involved in the production of the built environment. Built capital formation represents the total value of these activities. This focus on the totality of construction activities provides the proper basis for the formulation and assessment of policy, including policy directed at re-engineering or revaluing construction.

Definition of built capital

In the UK National Accounts, *gross fixed capital formation* measures spending on the production of new fixed capital in any given period, and includes spending on machines, computers, factory buildings and housing units. The construction of the built environment is an act of investment termed *fixed built capital formation*, which refers to the production of new buildings and other structures, and is measured by the amount of final spending on those items. As such it represents the broadest measure of construction activity, capturing the value added by all the different stages of construction – not just the stage represented by the 'construction industry'. Unlike the traditional definition of construction output, this definition captures all the stages of production; and measures of the value of this activity should not, in principle, vary with changes in the structure of the construction process. Total investment spending is termed 'gross investment' or 'gross capital formation'. Gross investment minus depreciation is net investment, and positive net investment increases the economy's total stock of building units and infrastructure. The formation of the built environment is a large and volatile component of national economic activity.

Measurement of the building stock

In national accounting terms, the gross stock of dwellings, other buildings and structures is valued at the actual or estimated current purchasers' prices for new assets of the same type, irrespective of the age of the assets. Capital stock data tend to follow a relatively smooth trend over time because they are recorded either as accumulated balances or as averages. Taking the UK as an example, Figure 10.6 illustrates this by showing the real net capital stock and net built capital stock for the UK for the period 1948–2003.

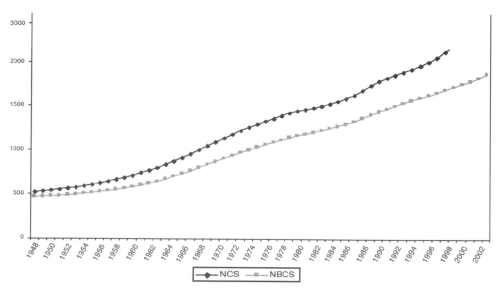

Figure 10.6 Percentage of net built capital stock (NBCS)/net capital stock (NCS).

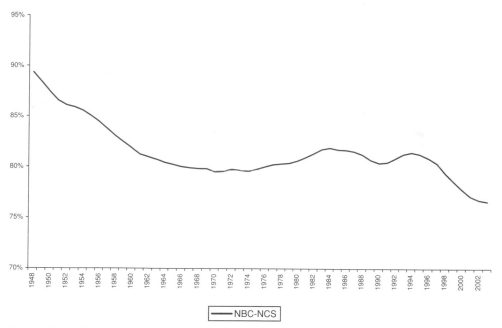

Figure 10.7 The percentage of built capital to capital.

Nowadays, the value of built capital accounts for a major part of the value of the UK's physical capital (physical wealth) – roughly 75% in 2003 – but this represents a substantial reduction in comparison with the post-war high of 89% in 1948. As can be seen in Figure 10.7, which shows the value of net built capital as a percentage of

Table 10.2 Value of stock (billion).

Building stock	Australia (2000) A$	Canada (2001) Can$	Denmark (2000) £	France (2000) £	Germany (2000) £	Lithuania (1997) Litas	Portugal (2001) £	Sweden (2000) £	UK (2001) £
Residential	582	953	242	2979	4 183	74	256	222	2266
Non-residential, civil engineering	720	953	105	1523	3810	35	138	143	1348
Ratio of stock value to GDP	2.0	2.0	1.9	3.2	4.0	2.8	3.5	1.6	2.4

(Source: Carassus, 2004)

78

the value of net capital stock for the period 1948–2003, the fall was not continuous. This suggests significant differences in the relationship between real GDP growth and built capital formation, on the one hand, and real GDP growth and non-built capital formation on the other. In times of steady real GDP growth, the growth in the value of built capital has been much slower than the growth in the value of non-built capital. Yet, during periods of pronounced business cycles, there are significant variations in the relative value of built capital, but not obviously pro- or counter-cyclical.

Value of the building stock

Built wealth accounts for roughly 20% of the world's total wealth (World Bank estimates, in Pearce, 2003). The average age of the stock is increasing – that is, the rate of depreciation of existing stock is low. Long-term changes in the real value of stock relative to real GDP indicate a structural change in the built capital : output ratio, and suggest changes in the productivity levels of built environment assets. The study undertaken by Carassus (2004) determined that considerable variation existed amongst the nine countries in the study (Table 10.2). The study concluded that, for housing, the value of the existing stock is easy to assess but non-residential buildings and civil engineering infrastructures are difficult to evaluate and are often underestimated. Estimation methods are not the same from one country to another, which makes direct comparison more problematical. The range of values for total construction stock varied from 1.6 times to 4 times the individual country's GDP.

The economic value of the built environment: broader issues

Urbanisation

A revaluation of construction is necessary in the broader context of growth of the urban environment. Continuous migration of people from rural to urban areas over many decades has led to a growth in urban populations in all parts of the world. As Figure 10.8 indicates, the rate of growth of urban population has, in recent years, been greatest in the poorer countries of the world; more people live in urban areas in low-income countries than in high-income countries, with the cities of Sub-Saharan Africa and East Asia showing the highest growth rates (Figure 10.9). The urban population is growing fastest in low- and lower-middle-income countries and the countries of Latin America had become as urban as the average high-income country in 2002.

Accelerating urbanisation presents both a need and an opportunity for the construction industry's contribution to be better understood. The increased construction activity has implications at both the local and national economy levels.

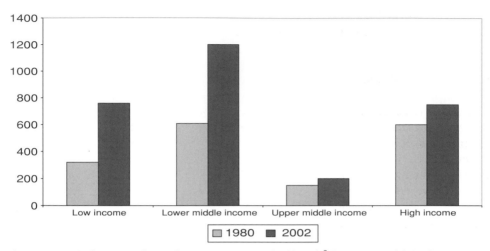

Figure 10.8   Urban population by income group (millions).[3] Source: World Bank, 2004a.

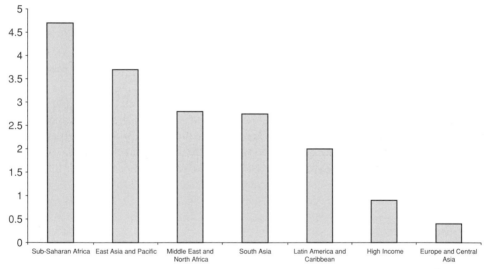

Figure 10.9   Average annual growth rate of urban population 1990–2003 (%).[4] Source: World Bank, 2005.

Environmental and social issues

As Pearce (2003) indicates, a true measure of the value of construction can only be calculated by broadening the traditional measures of value to include environmental

---

[3] For Figures 10.8 and 10.9, urban population is the mid-year population of areas defined as urban in each country and reported to the UN.

[4] Country status, in terms of income level, is as defined in: *World Development Indicators* (World Bank, 2005).

and social impacts. This should include such issues as: the level of investment in the housing stock; and a resource analysis of construction.

The level of investment in the housing stock

A significant fraction of the housing stock in the UK is in a state of poor maintenance and repair. Much of it is old and thus technologically inferior to new modern buildings, in terms of energy-efficiency, productivity (ill-suited to modern needs), etc. Spending on new houses is large and volatile. It therefore exerts a major impact on the economy. Most houses are purchased with money that is borrowed by means of a mortgage and, in many developed countries, interest on the borrowed money typically accounts for over one-half of the purchaser's annual mortgage payments. Variations in real interest rates therefore exert a substantial effect on the demand for new housing. Spending for new residential construction tends to vary negatively with real interest rates.

Resource analysis of construction

Using the UK to exemplify a typical developed economy, the use of existing building units accounts for one-half of the UK's final energy consumption and one-half of its carbon dioxide emissions. The (narrowly defined) UK construction industry has a conversion efficiency of 75%. Some 50% of the materials 'waste' is recycled. Much of the rest is inert. The industry (including quarrying, transportation of construction products and transport of waste) accounts for 5% of the UK's final energy consumption, and 30% of its industrial energy consumption. Mineral extraction and product manufacture account for 50% of the industry's energy consumption, and transportation accounts for 39%. Construction and related industries account for around 2% of all UK greenhouse gas emissions and 2–4% of other air pollution (Smith et al., 2002).

Conclusions

This chapter has attempted to put forward a case for a re-evaluation of the construction sector in order to deliver a step change in quality of output and thus develop a sustainable built environment. At the core of the discussion is the issue of the nature of a sustainable built environment – balance of new and old; conversion versus demolition. It is beyond the scope of this chapter to plan the journey required to transform the built environment from its actual state to a sustainable state. However, the perceived economic value of the built environment must be appreciated in order to properly evaluate the sector's full contribution to the quality of life.

It is worth reiterating the limitations of existing data sources, which prevent a true assessment of the value of construction activity. As suggested in a CIB Task Group survey report on sources of macro- and market data in construction (Ruddock, 2000), improvements to existing national statistical reporting systems are needed in many countries to ensure that all construction activity is measured. A fundamental principle of sources of data on construction activity produced by national statistical offices is that official statistics are essential for obtaining a transparent picture of the true value of construction. In spite of the UN's attempts to impose standardisation on national statistical bodies, there are still many differences in underlying concepts and definitions amongst different countries and the problem of coverage to incorporate the 'grey' construction sector is, in many countries, a significant issue.

The importance of relevant data for assessing such effects is of paramount importance and the Government cannot be relied upon as the only source of useful information. Cannon (1994) points out that many industry users may be unaware of the availability of some information because the originating institution may not recognise the usefulness of its dissemination to a wider audience. Official government statistics are not, of course, the only source of data and, in the industrially developed world, there is a plethora of private organisations producing data on the national construction sector. These tend to be forecasting institutions, contractors' organisations or information service providers. In those countries where a comprehensive information system does not exist, the CIB Task Group survey has shown that various options – such as government financing through a public/private agency, a subscription applicable to users or a levy system on members of the industry – could be considered. Improved statistical information would lead to improved awareness of the true value of construction activity. The target audience for this output would, of course, include government agencies for policy and planning activities. Additionally, Snyman (1999) identified four interested groups:

- Contractors, for information relating to the workload of the industry
- Professionals, for activity planning
- Construction materials manufacturers
- Clients

Finally, as the problems associated with the development of valid data on construction and the built environment tend to be dealt with on a national basis, but are, nevertheless, international in their nature, the value of information sharing and improvement might be tackled globally. Referring to the challenges facing information agencies generally, Keuning (2000) indicated that an international network of statistical agencies sharing information is needed to undertake data collection and improve its analysis. This would be, perhaps, particularly pertinent in the context of improving information on the construction sector and enhancing its revaluation.

11 Competing revaluing construction paradigms in practice

George Ang

Summary

In most countries of the world, the building and construction sector is a significant component of the national economy, with turnovers of 7–12% of GNP. However, this industry has been criticised worldwide for not sufficiently achieving the level of improvement in performance and productivity shown by other industrial sectors. This has created an internationally growing need for reform, which, in a few countries, is being accelerated by findings about systemic problems, including irregular pricing practices, artificial constraints on markets and even a degree of fraud. The urgent need for reform of the sector is apparent, and in a number of countries it has been initiated with the UK Rethinking Construction programme as leading principle. It is internationally recognised that reform requires a fundamental mind shift and a long-term time frame – a new vision for construction. Thus, this chapter looks in more detail at some of the issues raised in Chapter 4.

The central issue is what form of competition policy will enhance the right building and construction industry dynamics, and create market structures and competitive pressures that will drive reform and the revaluing of construction. The analysis in this chapter focuses on a few significant competing revaluing construction paradigms in practice, in an attempt to sort out the adversarial and somewhat conflicting theories and practices related to this central issue. In doing so the dilemma of 'innovation through competition versus cooperation' is considered a major discussion issue.

The dualistic role of government – on the one hand, balancing innovation and anti-corruption policies as a legislator and, on the other hand, acting as the prime client with a decisive role in public procurement – is recognised as an essential phenomenon in making reform happen. These roles must be represented separately in the process towards reform. Within this context it is argued that the development and application of alternative non-price procurement approaches is evident, because collusion will be sustained by focusing on lowest price only. A significant future trend is recognition of an explicit role of clients and a focus on end-users and, as a consequence, the need for best value procurement of the built asset as a whole is therefore apparent. This, in turn, calls for innovation through more cooperation, because such an approach drives prime attention towards whole building performance rather than the technology of its parts and components. In doing so, managing added value is explicitly addressed in the interrelation of product, process and people.

Interdependency, certain forms of cooperation and networking are considered to be in the nature of the building and construction process. This is revealed by evidence from different points of view in organisation economics and from findings of an international reform benchmark carried out in 2004 (Courtney et al., 2004). Moreover, the interdependency is in the need for developing models of process design and management, intrinsically based on 'multi-stakeholders value'. The adage 'competition is good, more competition is better' from other industries seems therefore to be oversimplified when it comes to the building and construction industry.

Finally, reform must be seen in a long-term framework in order to develop sound policies to make things happen, which initially need a political push but in the long run must be 'self-sustaining' and driven by commercial incentives. Innovation requires adequate support of research and knowledge infrastructure. Central to this view is a clear and well-founded vision shared by all significant stakeholders within the building and construction sector, and implemented in harmony with the specific culture of the involved nation.

Introduction

Within the perspective of revaluing construction, the phenomenon of competing construction paradigms is manifest and may be monitored and managed carefully in reform processes worldwide. To stimulate a challenging discussion, this chapter will highlight several major competing themes: commercial pressures from competition versus trust; cost versus value; and the need for political push towards reform versus its unreliability. Trust can be interpreted as being related to confidence or to cartels; the analysis in this chapter refers to the interpretation related to confidence, i.e. mutual trust on behalf of improved cooperation. The synthesis describes the nature of these competing construction paradigms and their possible impact on successful reform, the evidence mainly building on the findings from: the PSIBouw PP1 report on *Inventory of International Reforms in Building and Construction* (Ang et al., 2004), and from *Competitive Relationships in Construction – Competing with the Same Counterpart* (Doree et al., 2003).

The PSIBouw PP1 report (Ang et al., 2004) carries the results of an international benchmarking research initiative, designed as the 'pathfinder project' and instituted to provide guidance for other research projects within the PSIBouw initiative. (PSIBouw stands for the 2003–2007 Dutch national research programme on *Process and System Innovation in Building and Construction*, which was established in 2003 as a joint initiative between industry, government and research institutes.) The initiative follows the guiding principles of the 1998 UK *Rethinking Construction* report. The PSIBouw research programme supports the national *Action Agenda for Reform* of the Dutch building and construction sector, issued in November 2003 as a response to a Parliamentary Inquiry into large-scale fraud in the building and construction industry. PSIBouw has secured substantial funding from both government and industry.

In industrial organisation economics, positive market dynamics are seen as the driving force for progress (Schumpeter, 1949). Positive market dynamics ensure a sequence of business cycles in which organisations: (1) improve existing products and processes; and (2) in parallel, develop and introduce new products and processes. Each cycle has the following phases: introduction, expansion, maturity and stagnation (decline). Each phase has specific characteristics in terms of structure, behaviour and performance (Bain, 1956).

Within this context, competing revaluing construction paradigms are manifest in practice. Three main themes are highlighted in this synthesis:

- Commercial pressures from competition versus trust (with an emphasis on procurement)
- Value versus costs (with an emphasis on the role of clients)
- The need for political push towards reform versus its unreliability

Finally, an effort is made to indicate significant aspects of the interrelations between these three main themes.

Commercial pressures from competition versus trust

Findings and analysis from market dynamics inform the paradigm of competition and the commercial pressures related to it. This paradigm seems to be coloured by adversarial and somewhat conflicting theories related to competition versus cooperation. This makes up the background for the role of trust within the dynamics of building and construction.

On the one hand, commercial pressures from competition are manifest. To be competitive, market dynamics force suppliers to focus on customer needs and to continuously offer improved solutions against competitive prices. In an ideal situation, market dynamics encourage organisations to compete in these development cycles and to make progress. But, in reality, perfect competition and entirely free markets do not exist in most West European countries. This is because, in their neo-liberal market economies, governments regulate and supervise demand and supply processes in order to stimulate economic growth. In drafting competition policies, governments have to balance innovation and anti-corruption policies (Doree et al., 2003).

On the other hand, there is the need for trust in the relationship between demand and supply parties, with regard to both construction client and vendors, as well as along the supply chain, so leading to improved cooperation. This drives the notion that industry reform rests on creating more cooperation and less competition. But, there are attendant problems with respect to innovation, collusion, competition, concentration and anti-corruption actions. These problems appear to be symptoms of a stagnant market, which are most visible in market segments where public tendering and selection are dominated by the lowest price, in particular in traditional procurement approaches. Balancing the need for trust and commercial pressure from competition depends on creating the right industry dynamics for a particular country, culture and context.

Thus, a better understanding of the business systems and market structures, and of lessons learnt abroad, may substantially improve the chances of successful reform of this troublesome industry. The historical and international contexts show a consistent move away from traditional contracting. Industry reform typically rests on more co-operation and less competition. To an extent, this is an acknowledgement that the neoclassical notion 'competition is good, more competition is better', is unfit for the construction industry. However, sceptics doubt whether increased cooperation (i.e. better understanding, synergy and trust) will change the construction industry for the better. They argue that, in neo-liberal economies, markets and competition are the prime driving forces of innovation and growth. Cooperation, in their eyes, goes against healthy market dynamics.

Amongst policymakers, the ethos 'competition is good' seems to rule (Audretsch et al., 2001). The central idea of competition policies is to promote rivalry and avoid monopolies (Audretsch et al., 2001). In the construction sector, this delicate task is complicated by the fact that governmental organisations are themselves dominant market players (Doree et al., 2003). Recent Dutch policies clearly favour competition over cooperation.

So the issue becomes one of how to reconcile these apparently opposed views. What happens in the construction industry can serve as an illustration for a more generic debate on competition policy: what form of competition policy is needed to create the right industry dynamics (Doree et al., 2003)? The reform process must be based on a vision of creating market structures and competitive pressures that will drive reform and improvement after the initial impetus has faded. The process of reform will take longer than the lifetime of one Parliament, and to be self-sustaining must, in the end, be based on commercial incentives (Ang et al., 2004). These statements may be supported by the following international practices:

New South Wales, **Australia** – By 2000, some 59 actions had been implemented in New South Wales (NSW), Australia, and a further 18 were in progress. The next step in the evolution of reform was to encourage and facilitate greater industry leadership, in order to make reform self-sustaining. The Government established the 'Construction Improvement Roundtable' to provide a forum for the exchange of ideas on key issues affecting construction and to provide leadership in the creation of learning networks. The members of the Roundtable were appointed in their personal capacities. Members of the learning networks each make a formal commitment through a memorandum and then develop a business plan for the network. The networks enable members to benchmark performance, devise and use best practice processes and to participate in research. The CIDA's mission, to be a catalyst for real and measured change in the NSW building and construction industry, would be through the provision of leadership, mo-tivation and the development of a culture of learning and continual improvement. By setting up challenging performance standards, undertaking consultation with stake-holders to identify factors for success and remove barriers to change, they aimed to make the dynamic change process in the industry self-sustaining. Governments and industry should collaborate in the development of quality management programmes and systems with the aim of building a culture conducive to self-regulation.

Singapore – The professional (Singapore Institute of Architects, Institution of En-gineers Singapore, etc.) and trade bodies (Singapore Contractors Association Limited, Real Estate Developers Association Singapore, etc.) in this country are encouraged to promote continuing professional development (CPD) programmes for their members, and to make CPD programmes mandatory for the renewal of professional membership.

This is to cultivate a habit of lifelong learning amongst the industry players through a self-regulatory approach www.mom.gov.sg. The professional and trade bodies have also been urged to develop their respective codes of conduct, if they have not already done so. This would ensure that the industry players abide by a set of codes of practice in conducting their business and attain a high standard of practice. This is a first step towards self-regulation, image improvement and higher professionalism among industry players.

Hong Kong – The review committee has envisaged that, based on discussions and consensus building, the proposed co-ordinating body will have an active role to play in, at least, the following areas: to carry out self-regulatory functions for the industry through the formulation of codes of conduct and the administration of registration schemes for construction workers, subcontractors, renovation contractors and decorators, and other types of construction personnel.

The vision of industry dynamics and the related competition policy must therefore be part of a national vision on reform, which is then turned into reality through a long-term strategy. As with codes of practice, the development, promotion and implementation of a clear and shared vision can be a means of restoring and sustaining trust and creating confidence (Ang et al., 2004, pp. 45 and 56). For reform to be successful, the aims must be accepted by all the key stakeholders. No significant interest should be excluded from the development of the strategy or the subsequent reform process. The experience of Australia is illuminating: here, a national programme was developed without due input from the architectural profession and product suppliers, and, as a result, it had difficulty in securing acceptance. Established specialisms may appear threatened by new ways of working, and will need to be reassured. Even then, it will take time for new roles to develop (Ang et al., 2004).

Recent Dutch experience

Evidence suggests that the traditional competitive approaches, which most countries are now moving away from, give rise to systemic problems with market dynamics in the construction sector. It is argued that these traditional competition policies, due to the characteristics of the construction industry, create a business environment that encumbers innovation and dynamic efficiency. The construction industry, in particular the public works segment, seems highly vulnerable to ruinous competition (Doree et al., 2003). Traditional procurement approaches make it difficult for companies to start up new business cycles. Downturns in market volume induce destructive price wars and also exclude any form of trust, which was echoed in a collusion and fraud scandal in The Netherlands during 2002.

The Dutch Parliamentary Inquiry into the 2002 fraud scandal identified the following shortcomings in the building and construction industry that contribute to its relatively poor performance:

- Traditional market structures, with their emphasis on short-term relationships, provide little opportunity for optimising the relationship between price and quality or for continuous learning by both clients and the supply side.

- There is inadequate understanding of clients' real requirements and of the needs of society. To meet with these requirements and needs, the whole life-cycle of built assets should be taken into consideration. However, the traditional construction process is focused on the internal optimisation of sub-projects, and separates responsibilities for design, construction, operation, etc.
- The industry is highly fragmented, with many parties involved in the different phases of a construction project and, as a consequence, potentially subject to conflicting objectives and poor communications.
- Its research base and the bodies that advise on future opportunities and trends are not well linked to practitioners in the industry.

These findings complement those of a preparatory investigation, which also identified other factors that affect the industry's performance (Ang et al., 2004, p. 14):

- Clients have specific requirements and each building or structure occupies a unique site. Moreover, most production processes are site-based. This leads to one-off designs and ever-changing assembly processes. It is therefore difficult to compare the quality and price of different construction outputs, and this inhibits competition based on performance.
- In some markets, competition is reduced. Factors that have led to this situation include mergers, the cost of preparing bids and the criteria used for the pre-qualification of suppliers, which raise the threshold for entry.
- Firms have collaborated in tenders in order to apportion work, and this has again reduced competition.
- The strong 'horizontal' structures on the supply side, with separate, well-established bodies that represent the interests of architects, design consultants, contractors and product suppliers. These inhibit 'vertical' integration and are a barrier to improvement in the performance of the total supply chain.

These factors, it was claimed, led to abuses in the form of collusion on pricing and the allocation of work. They also inhibited competition and innovation, and so reduced progress in quality standards and productivity. There were inadequate incentives for higher performance or better value, and firms were not sufficiently orientated towards their clients. The overall effect was to give the industry an increasingly poor image, which put off talented young people from seeking employment in the building and construction industry. With an annual turnover of approximately £60 billion and comprising 85 000 firms and around 526 000 employees, the sector represents more than 7% of the Dutch GNP. The previous reports and the Parliamentary Inquiry therefore set the background for reform.

Due to market irregularities, and as a consequence of fraud and collusion scandals in a few countries worldwide, the restoration of trust has become a major issue in reform. In The Netherlands, political commitment to reform, now evident following the 2002 Parliamentary Inquiry on large-scale fraud, has been essential for the initiation of a national reform process in building and construction. Three Ministries (Trade & Industry; Transport & Civil Works; Housing, Spatial Planning and the Environment) issued an *Action Agenda* in November 2003, based on five main objectives (Perspectief voor de Bouw, 2003):

- Restoring trust between the Government and the sector
- Developing effective markets and a properly functioning sector
- Enhancing professionalism in procurement
- Instilling high standards in the supply chain
- Less, but more effective, regulation

A clear and well-founded vision has formed the basis of reforms in other countries. The process of creating this vision and the subsequent commitment is a contribution to restoring trust.

Codes of Practice and Codes of Ethics appear to be valuable tools for the restoration of trust and the establishment of proper relationships between clients and supply-side interests. These tools should also be adopted within the supply side in order to provide a framework for commercial relationships down the supply chain. Acceptance of such codes by supply interests signifies a commitment to working to high principles, with integrity in all transactions and respect for staff, clients and partners in the supply chain. However, to be effective, codes need to be incorporated in commercial processes (e.g. registration or pre-qualification schemes). Compliance with a code should be monitored and sanctions applied if that code is breached. In this way, the codes become part of the market pressures for reform and improvement.

Such codes have normally resulted from a joint initiative of government and industry, but the example of New South Wales shows that initially they can be imposed unilaterally through collective action by powerful clients (i.e. government departments). After a period, however, an amended code was developed in NSW by collaboration between NSW government and industry.

In The Netherlands, therefore, issues to be considered include not only the role of codes of practice in restoring trust, but also the strategy to be adopted in their development and application. Some construction interests, notably in the design professions, have existing codes that cover their responsibilities as professionals. These may need to be re-examined as part of the reform process, since the experience of other countries is that professionals may be inhibited from participating fully in reform either by traditional codes or by statutory restrictions.

So, there is evidence of good experience in the use of codes of practice or of ethics as a means of restoring trust between the government and the industry and among parties in the supply chain. The development of such codes is normally undertaken through collaboration between industry and government and this, as well as the ultimate commitment of all parties to the code, contributes substantially to strengthening trust between them (Ang et al., 2004, p. 55). There is, though, a feeling that codes may secure decent, but not innovative behaviour.

The Dutch Advisory Board on Science and Technology reports (AWT, 1997), on the relationships between regulation, competition and innovation, point out that technology and competition policies must be in balance, but that it is difficult to indicate what type of rules are required. Here, two logics collide: on the one hand, competition and commercial pressures are seen as the key to innovation; on the other, collaboration and trust are seen as essential for innovation and technological breakthroughs (joint efforts, clusters, supply chain integration). Anti-corruption regulations and innovation policies often seek the same outcome with strategies that contradict each other. This is illustrated by international trends in procurement, in particular from the 'Hong Kong Case'.

International trends in procurement

Reform initiatives in several countries show that such reform requires committed national initiatives, including public sector procurement policies (Ang et al., 2004). The construction reform policies that have been adopted stress the movement away from adversarial relationships and lowest price selection (NAO, 2001; Strategic forum for construction, 2002; Revaluing Construction, 2003). An international trend can be distilled based on the proceedings of the *Revaluing Construction* conference. This trend is towards value and quality driven competition, integrated team delivery, long-term commitments, public–client leadership, development of benchmark instruments and joint initiatives by public agencies, private enterprises and universities. The international trend shows a consistent move towards non-price- and value-based procurement, away from focusing on lowest price only. But the interesting driver for reform may be in the response to the question: What form of competition policy is needed to create the right industry dynamics? The role of (public) procurement in reform is evident in Australia, Norway, Finland and Hong Kong.

The Hong Kong case is selected below for two reasons: (1) its dealing with the competition/cooperation paradigms, and (2) trust in terms of anti-corruption measurements (Ang et al., 2004, pp. 91–98). In response to corruption scandals during the 1980s and 1990s, political concerns over the industry's performance led, in 2000, to the establishment of a high-level commission, the Construction Industry Review Committee (CRIC). This committee was chaired by the present Minister of Finance, and with membership drawn from the construction and property sectors, trade unions, academia and government. The CRIC (2001) report *Construction for Excellence* was implemented with procurement at the heart of the reform strategy.

Many of the CRIC recommendations were directed to public bodies in their role as clients. They have introduced new procedures as a consequence, but the impact on industry performance, other than on safety, is not yet demonstrable through statistics. Major infrastructure projects, which have involved many international consultancies and contractors, have been procured and delivered through processes that would stand comparison with anywhere in the world. Measures introduced in public sector procurement include:

- Greater emphasis on past performance in tender evaluations, and the introduction of non-price factors generally in tender evaluations
- Revised 'two-envelope' system for the selection of consultants
- The development of an ethics code of practice
- Establishment of a 'premier league' of contractors for strategic partnerships
- Greater sharing of performance information amongst public clients
- The successful use of dispute resolution advisors

With regard to the theme of trust, an anti-corruption policy was implemented very explicitly in the establishment of the Independent Commission Against Corruption

(ICAC). And it is significant that this ICAC, a highly influential body in Hong Kong, is relaxed about new procurement methods. They do not regard them as more prone to corruption, provided the rules are transparent and the procedures followed. But next to the core business of preventing and prosecuting corruption, the ICAC puts a lot of effort into education and training in order to inculcate the right attitudes. Points for particular international consideration include the following:

- The political drive for reform was reflected in the industry, which welcomed the focus on its contribution and issues provided by the CRIC. It also brought to light many examples of good practice, as a counter to criticism of the industry. There was therefore a willingness along the supply chain to support the Government initiative and to accept the Committee's recommendations.
- CRIC had representation from all significant interests (although its members were not nominees of particular bodies), which also assisted acceptance of its recommendations.
- Procurement reform had a central role, with key public bodies not only reviewing their own procedures, but acting as a focus for broader government initiatives. Specific tools, such as the Performance Assessment Scoring System of the Housing Authority, deserve further study as do systems employed in Hong Kong for dispute mediation, for registration of contractors and sub-contractors and for pre-qualification of contractors and consultants.
- The CRIC review was very broad. There is a risk of loss of focus, particularly if the broad strategy is not articulated and constantly promoted. Consideration of the scope of reform should be combined with assessment of the resources available for promoting industry debate and change, through supply-side initiatives, not just procurement reform.
- Highly effective procurement practice exists alongside poor practice, but in Hong Kong the gap between the different classes has not been bridged. Performance data from the best public and private projects, collected through normal monitoring procedures, could be used to stimulate change elsewhere. This would require leading firms and clients to be exemplars, and the development of performance indicators that are widely applicable. However, the business drivers in parts of the construction market may continue, as in Hong Kong, to favour short-term perspectives.

This case may also illustrate the point that the strategic objectives of reform may vary. In Hong Kong, the programme aims essentially to improve the industry whilst retaining its present structures of responsibility – professionals, contractors, etc. But more radical change could have been contemplated. The experiences from Australia, Norway, Finland and Hong Kong (Ang et al., 2004) reveal that the role of public procurement is certainly a key means of developing an effective market; however, these reforms illustrate some different aspects, such as:

- Adoption of more integrated approaches to procurement
- Inclusion of non-price factors in tender evaluations
- Adoption of life-cycle costing or moving in the direction of 'concession', i.e. design–build–operate contracts

- Selection of consortia for 'programmes' of projects, rather than a single project
- Development or revision of pre-qualification schemes
- Collection (and possibly publication) of formal project monitoring data
- Development of shared benchmarking databases

It is the experience in these countries that, while public sector procurement policies can be co-ordinated, this is much more difficult in the private sector. However, there are private sector clients with large property estates, and these should be included in the reform process if at all possible. The development of codes of practice, national registration systems for construction firms and benchmark indicators, may be one way of encouraging different procurement routes in the private sector. A major issue is to what extent private sector procurement can be aligned with those adopted by the public sector, in order to strengthen the market pressures for reform.

None of the possibilities discussed above is likely to be appropriate universally; further study will be required to establish the optimal combination of tools for different types of projects. The creation of a forum in which these issues can be considered by public and private sector clients, and changes agreed with the bodies responsible for auditing public expenditures, appears to be an early requirement. This is discussed further in the next section.

Last but not least, one should be aware of the fact that there is no evidence that quality and innovation are the results of a good process alone. The issue of value in product, process and people must therefore be addressed more explicitly than in traditional procurement approaches. This issue will also be treated in the following part, where the focus is on the competing paradigms of value versus costs.

Value versus costs

The design of public buildings, it is said nowadays, must have a measurable effect upon staff, end-users and relations with the local community (HMSO, 1994, p. 66). For part of the function of government buildings is to arrange for 'an overall sense of employee and user-wellness, both physical and mental'. Design, it is 'officially' believed, can perform astonishing feats of social engineering and 'cultural change'. Government buildings may pioneer design as social engineering. This illustrates the scope and ambition of the 'value' perspective.

However, the terminology used within the context of value does vary depending on the level of stakeholders' interests. On the level of the workplace, fitness for purpose is an important issue; while on the level of asset management, return on investment is certainly the decisive issue; and on the level of society, there may be the issue of value in terms of cultural, sustainable and architectural quality. The general approach, though, is based on the assumption that, in client expectations of total building performance, the quality and yearly costs of the workplace will become identified as a key to business competitiveness. In many countries programmes are ongoing with the aim for more or less radical reforms in the building and construction sector in order to 'value-empower' the sector, and procurement appears to be a key factor for improvement in terms of enhancing the delivery of added value and client satisfaction.

International spearheads are recognised in procurement as the key towards reform, in codes of practice and integrity, resulting in sound options for pre-qualification through registration of past performances. Best value procurement is the leading ambition. The price : quality ratio of the output should drive the selection of vendors rather than the focus being on the lowest tender price alone. In doing so, best value procurement may rely on the following objectives (Kashiwagi & Verdini, 2004):

- Increase the efficiency of operations
- Minimise risk of non-performance of outsourced services (hire the vendor who is the most efficient, put the vendor with the risk and the risk will be minimised)
- Maximise contractor profit while delivering best value to customer (lowest price and best performance, maximise the contractor's profit by being efficient on the client's side)
- Minimise the need to manage outsourced and in-house services

The increasing importance of end-users' satisfaction creates the need for a performance-based approach to allow the management of end-results, and to provide a language they can understand. With such an approach the match between the clients' business objectives and the buildings will be facilitated and, as a consequence, buildings can be treated as factors of business rather than a capital asset only. Within the perspective of reform, best value procurement seems to be a key factor, and in this respect the role of the construction client is expected to be decisive. A number of interesting views arising from the *PSIBouw PP1 International Reform Benchmarking* report may illustrate the impact of this client role (Ang et al., 2004).

Establishing client association 'clubs', where information can be exchanged and new ideas tested, is one valuable means of creating that confidence and generally stimulating a more professional approach to procurement. These may be associations of clients, as in Denmark, or groups of individuals, as in New South Wales. The Danish association arose as a result of a document called *Danish Construction Sector in the Future – from Tradition to Innovation* (Building Policy Task Force, 2000), issued by a task force appointed by the Government. The Australian council's membership ranges from every leading institutional investor, pension fund, property trust (Australia's REITs) and financial organisation to private investors and developers. Asset managers, the professions and trade suppliers actively participate in the association (www.propertyoz.com.au). There remains the case of one-off clients and, traditionally, professional advisors (often architects) and consultants help these clients to fulfil their role. But even if architects or consultants act on behalf of the client on the demand side, they belong to the supply side and cannot fulfil the need for exchanging concerns that are particularly related to demand and the client as an end-user. Among others, a specific foundation for individual house owners has been established in The Netherlands, which cares for the interests of these end-users.

Each of these mechanisms creates a community amongst clients engaged in procurement reform. Providing the opportunity for the identification of leading practices, joint examination of issues and learning how others have resolved problems, can be a means of strengthening that community (Ang et al., 2004). More ambitiously, such clubs can become engaged in research, particularly into new ways of identifying and expressing client needs. The development of a client-focused industry, widely advocated as a reform aim, will be promoted if clients can more precisely relate their construction requirements to their business and organisational needs, and if they have the means of monitoring subsequent performance. At present, the knowledge base required is inadequate, but the collective experience brought together in client clubs could inform the research required.

In cooperation between the Dutch Government Building Agency (VROM Rijksge-bouwendienst) and the International Council for Research and Innovation in Building and Construction (CIB) the *1st International Client Platform* was held on the 23rd and 24th September 2004 (ICCF, 2004). Professional construction clients from Australia, Canada, Denmark, Finland, Norway, South Africa, Sweden, the UK and the USA attended the conference. This first meeting agreed on the value of convening the international client platform at periodic intervals, hosted in turn by different countries. The South African Construction Industry Development Board (CIDB) hosted the 2005 international conference.

This sharing of good practices highlights the key role of clients in implementing a range of partnering models. In the Danish definition, for example, partnering is an innovative and efficient form of collaboration which is based on dialogue, openness, confidence and with early involvement of client and companies. The definition furthermore underlines that the project is implemented by common targets formulated by common activities and based on common economic interests. The basis for the partnering contract is a partnering agreement with ground rules for cooperation. All partners do commit to this agreement. The selection of partners is based on fixed criteria like past performances, financial figures, etc. From the 1st of January 2004 it has been compulsory in Denmark for government departments – in every new project – to evaluate the application of partnering principles in a systematic way and to document the results. Guidelines for partnering principles in practice have been issued.

The issues discussed above are important routes to changing relationships in general and procurement policies in particular, but there will be many challenges in their application. Experience from Australia and elsewhere suggests that it takes consistent effort over a period of years to achieve change in all levels of an organisation. This is not only a matter of communication, staff need to have the tools that will enable them to implement new approaches and to be confident that these are leading overall to better results, even if at times it appears that extra expenditure is being incurred.

The central place of procurement in reform leads naturally to a much greater focus on client requirements and on the need to raise their capabilities. Client clubs can identify and promote good practice and help to give staff confidence in new procedures. They can also participate in research. It was noted that even where reform has been in progress for some years, there were allegations that government clients were not consistently following the principles that they espoused, and this reinforces the need for support for staff in the move from price-based selection. Thus, client associations can help clients become more professional.

Phenomena in the value versus costs competing paradigms

The competition between the paradigms of value and costs in practice can be illustrated by a few phenomena, derived from the industry's struggle to start new business cycles (Doree et al., 2003):

- When the acknowledgement of value and reputation is not possible, cost control will become the dominant strategy. Resources are immediately idle. But a cost focus drives organisations towards lean cost structures ('anorexia'). Stock buffers are not an option to protect against upturns and downturns. A discontinuity in workload means that resources are immediately idle. Consequently, firms invest as little as possible in capital goods, outsource as much as possible, and share risks by working in consortia. To limit problems, and to control progress in projects, methods and materials are standardised. As a consequence, construction firms are almost interchangeable and price is the only way left to compete.

- Selection on price suppresses competition on quality. It assumes that all suppliers offer products of equal value, and ignores the fact that competing firms may be encouraged to find solutions that create higher value for customers. As a result, contractors do not develop sensitivity for the wishes and latent needs of the clients.

- The traditional 'design–bid–build' approach leaves little space for developing new products and technologies (AWT, 1997). Clients dictate solutions, and there is no real demand for creativity.

- The final consumer of a construction product is generally not involved in the transaction process, and therefore the added value to them is left out of the competition equation.

- Since no branding strategies exist, there is little possibility of translating reputation and goodwill into market power.

- Production-on-location hinders scale benefits and makes advance production (buffer stock) impossible.

- Reactive market positions lead to short-term strategies.

- Floating capacities cause bullwhip effects (when the market shrinks, orders will also shrink and, as a result, more firms will bid for the fewer available tenders. This reduces the likelihood of gaining an order, which leads to even more firms bidding for each tender, etc.).

The above mentioned phenomena may give cause to bring the issue of creating added value under more explicit attention. Added value can be seen in the interrelation of product, process and people. In the building and construction process, the products' added value is initiated by the architects' design, and will be manifest in the way elaboration is reinforced by the other parties' expertise, and finally in a sound post-occupancy evaluation result. From well-known practices, the cross-functional network team (also called the 'building team') can be considered the most advanced form of cross-functional integration (Halman & Prins, 1997, p. 4). An increased use of joint ventures, subcontracting and licensing activities occurring across international borders, and new business ventures spinning off from established corporations, are already evident in several industries. These network-based organisations will most likely have one or more of the following characteristics (Miles & Snow, 1986; 1997):

- Division of labour based on specialisation of functions. Business functions (such as product design and development, manufacturing, marketing and distribution) are performed by independent organisations within the network.

- *Linkages among partners.* These are created and often managed through a 'broker-like function'. Besides a co-ordinating role, a broker may also perform one of the specialised functions within the network.
- *Value added partnerships.* Financial compensations are especially based on agreements on the added value of each of the contributing partners.
- *Shared information systems.* Based on mutual trust, participants in the network organisation share their contributions and are open to an exchange of information in continuously updated information systems.

In the building and construction industry, these network-based organisations are manifest in, among others, the turnkey network team, the cross-functional network team, all kinds of partnering principles, and in new developments of architectural design management (Halman & Prins, 1997). There is, though, a major point of tension that may hamper networking, namely the strong 'horizontal' structures on the supply side, with separate, well-established bodies representing the interests of architects, design consultants, contractors and product suppliers. These inhibit 'vertical' integration and can be a barrier to improvement in the performance of the total supply chain.

The idea of value added partnerships therefore covers financial compensations between the contributing partners, based on the added value of each partner in the process, and may be the subject for prime attention in making this happen in daily practice. Complementary to this, the right team dynamics must be achieved. In order to achieve the team dynamics to 'make this happen', the Kolb Learning Cycle may be helpful for creating a common mindset (Barrett & Stanley, 1999, p. 75–76). One way to categorise individuals springs from their emphasis on different parts of the learning cycle:

- **Sensing:** observations and reflections
- **Watching:** formulation of abstract concepts and generalisations
- **Thinking:** testing implications of concepts in new situations
- **Doing:** concrete experience

From this it can be seen that architects tend to be sensors–doers; lighting/sound engineers watchers–thinkers; and quantity surveyors watchers–thinkers–doers. The important thing is that the whole learning cycle is represented in the design and construction (and eventually maintenance) team, given that it seems very unusual for any one individual to have a full spread. Thus, in parallel with ensuring that the necessary technical knowledge is accumulated within the team, complementary learning emphases are also desirable.

Basic to a dynamic control concept is the interactive mechanism of value demand and value supply as described by de Ridder (2002). This model, based on Porter's management theories (de Ridder, 2002), distinguishes between value demanding parties, striving for a maximum difference between value and price (i.e. added value/benefit) and value supplying parties striving for a maximum difference between price and costs (i.e. profit) (Figure 11.1).

The essence of the value–price–costs model is that price represents the tuning mechanism between demand and supply (de Ridder, 2001). Consequently, the resulting added value of a project will be of interest to both the demanding parties and the supplying parties. On the demand side the desirable value and price are reflected in the

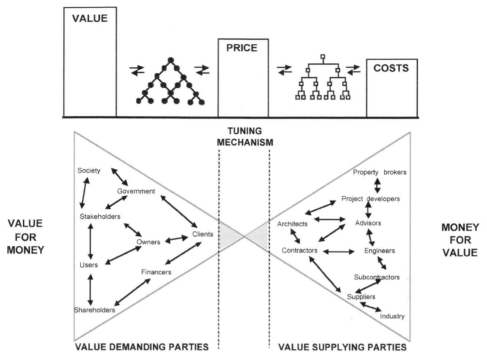

Figure 11.1 System model: network structure of value demanding and value supplying parties (de Ridder, 2003).

performance-based brief, and on the supply side the price and costs are reflected in the project concept, logistics, risk analysis and calculations.

This complex relation between value, price and cost is depicted by treating the price as a simplified tuning mechanism between the desirable demand and the possible supply. In the next sections the issues of value, price and costs are decomposed in detail, to properly explain the benefits of the interaction of value, price and costs. In order to monitor and sustain value across the process from project initiation to hand-over, use and maintenance, three domains are described. These are characterised by activities and interests related to demand, production and use. In addition, points of transactions and risks are recognised within these domains.

As an example, the dynamic control concept is applied in a performance-based building process (Ang et al., 2003). A practical approach of the integral treatment in performance-based building is to treat the business of the demanding parties/users separately from the business of the producing/supplying parties. Consequently, three domains can be identified, i.e. the domains of demand, production and use/facility-management (Figure 11.2). Within a performance-based building process, or other design & build-like organisation forms, two main transaction points can be distinguished. The first one is between the briefing and design phase, at the shift between the domains of demand and production. The second one is between the construction phase and the delivery of the building, at the shift between the domains of production and use. Now, the critical success factor in realising value within this approach is in treating the transactions as a dualistic negotiating process and not as a linear delegated

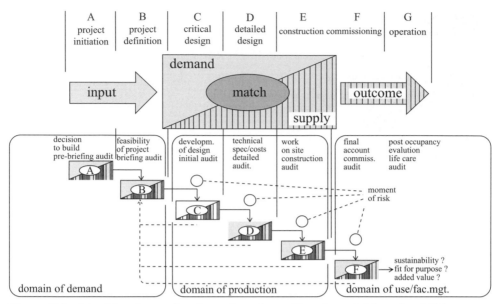

Figure 11.2   Matching demand and supply and specific points of risk at the shift between phases in the production domain.

package of risks thrown on the contracting partners, which traditionally is the case.

The mechanism of value demand and value supply, i.e. balancing value and price versus price and costs, should continuously be manifest in all phases of the process as a dynamic performance-based control concept (see Figure 11.2). The points of risk are a consequence of the sub-processes of matching demand and supply, whereas quality and value are expressed in a different way in each phase. However, the shifts between phases within the production domain introduce both specific points of risks (Ang et al., 2001) and opportunities for mutual gain when the application of the dynamic control concept can lead to alterations in the design & build contract. The design & build process, by its very nature, is especially characterised by unforeseen events, which may frustrate progress. The points of risk are a consequence of the sub-processes of matching demand and supply, whereas quality and value are expressed in a different way in each phase.

The proposed dynamic control concept allows the demand side to set new goals when confronted with an undesired and unexpected event, provided the supply side is compensated financially when more added value is created. In the next section, the concept of architectural value will first be discussed and defined. Then a dynamic control model will be developed to provide better prospects for managing and safe-guarding the added value of architecture.

The basis for dynamic project control is a simple method of value, price and costs. Two basic strategies can be distinguished to create maximum benefit (Figure 11.3):

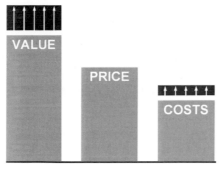

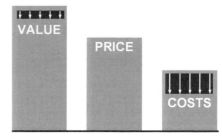

Strategy 1: Extra value versus limited costs

Strategy 2: Less value versus less costs

Figure 11.3 Two strategies in value–price–costs.

- The first strategy is to add a substantial amount of extra value against a limited amount of extra costs.
- The second strategy is to accept a limited amount of less value against a substantial amount of fewer costs.

Summary from a Dutch perspective

More integrated forms of procurement have been found to provide better value for clients. Having greater integration of design teams, and extending integration to include construction, has resulted in higher quality construction and more effective processes. The Dutch building and construction sector is fragmented and, while this partly reflects the wide diversity of activities and the variation in the scale of projects, it also leads to poor communications and inefficient processes. The reform programme in The Netherlands should, as a key issue, promote appropriate integration in order to address fragmentation of the process. In the long term, reform principles are embedded in the industry through training and through the education of future entrants, and reform processes should therefore involve the educational system. In Singapore, multi-disciplinary courses for professionals have been developed to stimulate mutual understanding and integrated working.

Procurement practices are of crucial significance in reform. Traditionally, procurement has been based on price. The introduction of 'non-price' factors in the award of contracts and of 'value for money' criteria and innovative approaches to tendering can have major impacts. Registration of potential suppliers, with past performance being one of the factors taken into account, can be the basis of a transparent and accountable system of pre-qualification, which also stimulates performance improvement. While procurement systems are very influential, it was widely accepted that successful projects resulted from teamwork and having the right relationships. Such relationships may be promoted by more integrated procurement systems, but there were also examples (e.g. in Norway and Australia) where teams were specifically assessed for the quality of the cooperation as part of the selection process. A focus on the factors that lead to good human and organisational relationships in construction should be part of the reform programme.

The need for political push towards reform versus its unreliability

There may be an urgent need for an initial political push, but to be sustainable, reform has ultimately to be driven by market forces. Change can be stimulated by specific actions, such as national political drivers, taking place against a background of concern over the performance of construction. But the time scale for reform is measured in years and possibly decades, and over that time political priorities change and the attention being given to construction following national political drivers will fade. Long-term change will take place only if markets provide rewards and incentives for superior performance and have within them mechanisms that stimulate continuous improvement.

The construction sector is large and diffuse. The client community is diverse, and many clients have little experience of dealing with construction. The industry has developed its practices over many years, and has developed many institutions and specialist groupings that reflect current ways of working. The experience of many countries is that changing attitudes and practices is a process that takes up to ten years, and even then there will be more to be achieved. Countries whose reform processes started four or five years ago may have implemented a good number of the immediate reform measures, but they also recognise that these have yet to have an impact throughout the sector.

For example, in Singapore, 'Construction 21' started in 1999, and is still in progress. New South Wales (Australia) pursued a reform strategy based on public procurement from 1992 to 2003, applying forms of pre-qualification based on past performances. The Queensland Government commenced its reform in 1994, and is still active in improving industry performance. In Denmark 'Project Hus' was planned to last from 2001 to 2009. In Finland, TEKES, an agency of the Ministry of Trade and Industry, has promoted industry change through its technology programmes for over ten years.

In The Netherlands, political commitment to reform, now evident following a 2002 Parliamentary Inquiry into large-scale fraud, has been essential for the initiation of reform, but is not the permanent driver that is required. There are important lessons from Australia and Denmark, where initiatives faded away within four years as a consequence of government changes. A few illustrations of this phenomenon come from Australia, Denmark and Singapore (Ang et al., 2004, pp. 8, 101, 121):

- **Australia: the demise of the CIDA** – The Construction Industry Development Agency, Commonwealth 1992–1995) was established with a 'sunset clause' that had the effect of bringing the Agency to an end after three years. During its period of operation, there was a change in government and even while the CIDA was functioning, there was a review of its activities. There was no political support for extending its period of operation.

- **Denmark: the termination of Project Hus** – The most recent reform programme in Denmark, 'Project Hus', was aimed at improving construction processes and had many similarities with the Dutch PSIBouw research programme. After the elections in Denmark in 2001, the new government took the view that clients and the market should be the principal forces for change and terminated Project Hus after two years.

- **Singapore: reform becomes routine** – While the Construction 21 reform process is still active in Singapore, some participants expressed the view that it was losing momentum and that the reform activities were becoming routine. Some new stimuli were thought to be needed.

These experiences suggest that a central aim in a reform process should be to create pressures for change and improvement that become part of the industry's normal practices, so that ordinary commercial competitive processes have the effect of stimulating industry improvement. Through such means, reform can be embedded in the industry and will survive political changes. The new procurement processes developed in Hong Kong and Australia – which use past performance as a factor in the selection of contractors, the registration of contractors as practised in Singapore, and the benchmarking being introduced in Denmark – are examples of changes that will apply continuing pressure for improvement to firms in the industry.

In a period of high political attention on construction, and a will on all sides to change attitudes and practices, there is an opportunity to introduce new procurement systems and other measures that will provide the necessary pressures and rewards. This is the opportunity to be addressed by high-level steering councils of reform, supported by national research programmes. The process of reform will therefore definitely take longer than the lifetime of one Parliament, and to be self-sustaining must in the end be based on informed (maybe a role for research and education in the broadest sense here?) commercial incentives. This theme runs through all the findings of the 2004 PSIBouw PP1 International Benchmark study (Ang et al., 2004). However, it is clear from international benchmarks that, with time, clear purpose and commitment, it is possible to change a situation characterised by suspicion, corruption and lack of trust into one of collaboration and mutual respect, with the reform strategy changing as relationships evolve. And that real improvement in industry performance can be secured through government action. These are encouraging conclusions for other countries dealing with similar concerns (Ang et al., 2004, p. 112).

The public sector is a very significant client for construction, and it is therefore essential that public sector procurement practices reflect the objectives of reform. Some reform programmes, for example in New South Wales, have been based on the power of public sector procurement. Audit bodies have supported reform (e.g. in Hong Kong) and should be involved in the development and implementation of new procedures. It is less easy to secure a coordinated approach from the private sector, but leading private sector clients in Hong Kong and Singapore have adopted the same procurement principles as the public sector and so reinforced the reform process. This should be encouraged in other countries with similar problems. But does the public sector really lead the private sector or follow it in practice?

While the circumstances that have stimulated the desire for reform in The Netherlands are unfortunate, they are not unique. It is encouraging that other countries, although they would admit that much remains to be done, have achieved considerable change in their building and construction sectors through reform processes. But their experience is also that reform does not happen overnight – it is a long-term process – and ultimate success requires constancy of purpose, the commitment of all key stakeholders and the development of business processes that stimulate and reward high levels of performance.

For reform to be successful, the aims must be accepted by all the key stakeholders. No significant interest should be excluded from the development of the strategy or the subsequent reform process. The experience of Australia is illuminating, where a national programme was developed without due input from the architectural profession and product suppliers, and had difficulty in securing acceptance as a result. Established specialisms may appear threatened by new ways of working and will need to be reassured. Even then, it will take time for new roles to develop.

Conclusions

The recent Dutch 2002 practice of collusion and, as a consequence, lack of trust may be a phenomenon incorporated within the structure of the sector. Characteristics that can be derived from this practice and contribute to collusion are manifest:

- In the need for collaboration so as to prevent discontinuity
- In securing that processes and products will not be subject to substantial change
- In creating barriers for access of new suppliers, i.e. 'outsiders'

The central idea of competition policies is based on competition in order to promote rivalry and avoid monopolies. But collusion is not the single root of the problems in the construction industry. The focus on lowest price procurement, traditionally chosen by decision-makers all over the world as an interpretation of this central idea of competition policies, provides an undesired effect in sustaining collusion. This effect springs from the fact that both the decision-maker's behaviour and the result of tendering are pretty much predictable in lowest price procurement. Industrial economics suggests that any segment of the construction industry dominated by a highly competitive price-driven public sector procurement approach will have a natural tendency to drift towards ruinous competition: a market state prone to concentration and collusion (Doree et al., 2003, p. 8).

This chapter describes the nature of competing revaluing construction paradigms and their possible impact on successful reform. The central theme in this reform seems to be determined by the search for the right competition policy, chosen to create the right building and construction industry dynamics in harmony with a nation's culture. Against this background a number of remarkable conflicting pressures are manifest, and need sound balancing and management.

Starting the debate over strategic objectives and the shape of the future industry could be an important element in a response to the prime question: What form of competition policy is needed to create the right industry dynamics? The 'neo-classical' notion 'competition is good, more competition is better' seems to be oversimplified and as such unfit for the building and construction industry. There is a certain interdependency between the parties in the building and construction supply chain that leads to networking. Since the quality : price ratio of a building and construction object is related to the built asset performance as a whole, rather than to an independent building part or component, the participants in the supply chain will be unable to individually deliver a sustainable result, unless all actions are subject to sound coordination. Therefore cooperation will always belong in the nature of this process, and it will be manifest in a culture-related way depending on the nation involved.

Added value requires an attitude of consideration that investments in time and money may be worth while, even if budgets are exceeded. It is an attitude and culture of thinking and doing that goes far beyond managing a proper building and construction process. With regard to adopting this attitude, many problems stem from 'problems of language' which originate from the parties within the process looking in a fragmented way at the nature of the process and, consequently, at the issue of value. This requires the establishment of shared values, which goes far beyond process, money and risks. Traditionally the primacy of architects is determined by the 'power of design', which can be seen as the first comprehensive prototype of the asset in use. Generally, in the phase of go/no go decision-making during project initiation, the other parties and their expertise are left out. Within the context of sound reform, the added value of the architectural concept will be manifest in the way the elaboration of this concept is reinforced by the other parties' expertise. This may provide conditions based on 'multi-stakeholders' value', in favour of a more balanced position of the architect in securing added value within the process.

The balance of conflicting pressures, between commercial incentives and trust, competition and cooperation, value and costs and, finally, the need for a political push versus its unreliability, should be managed through the establishment of a clear and well-founded vision. This should be developed as a joint vision of industry and government, and be shared by all significant stakeholders within the sector, so engendering trust and confidence among the parties and allowing progress to be made through a long-term strategy.

The process of creating this vision and the subsequent commitment among key stakeholders within the sector has been experienced as a major contribution to both sustaining – and eventually restoring – trust, and the success of reform. This also refers to the earlier mentioned statement that interdependency is in the nature of the building and construction process. The form of commercial pressure from competition policy and industry dynamics must therefore be incorporated in such a vision, and implementation may go hand in hand with the specific culture of the subject nation.

Governments have a dualistic role in the sector's network: on the one hand, as the legislator; and, on the other hand, as the public client in procuring public buildings and civil works. This is recognised in the Hong Kong case by the CIRC involving legislative and procuring roles separately in the CRIC. The lesson is that the government as legislator and as procurer must be represented separately in order to effect reform. Since collusion seems to be sustained by traditional procurement on lowest price only, it is argued that public procurement should focus on the introduction of non-price factors in procurement, and the use of registration of past performances and pre-qualification schemes.

The International Reform Benchmark underlines the role of (public) procurement. The public sector is highly significant in the overall market for construction, and it was evident in all countries that the policies and practices of public sector clients were key influences on the industry. While individual public bodies will have some special requirements specific to their functions, experience elsewhere is that they are able to adopt common principles in their procurement practices.

Historical and international contexts show a consistent reform trend from more co-operation and less competition. The Dutch practice teaches that only focusing on traditional procurement and price competition may not lead to the right building and construction industry dynamics. This is recognised in international trends consistently moving away from traditional contracting. The International Reform Benchmark study (Ang et al., 2004) illustrates a few alternatives. For example, the introduction of non-price factors in procurement, and the use of registration of past performances and pre-qualification schemes as developed in the PASS (Performance Assessment Scoring System) in Hong Kong and similar instruments from the Australian APCC (Australian Performance Construction Council).

Finally, Doree et al., (2003) recommend standing back to gain perspective, referring to Audretsch et al., (2001) and authors such as Schumpeter (1949), Shackle (1971), and Kirzner (1973). These theories urge us to see competition in a longer time frame, providing firms with the opportunity to constantly create new products and processes, and to start up new business cycles, in order to gain competitive advantage. Companies need brief periods of relative monopoly to gain the incomes to fuel this process of renewal. Short-sighted competition, aimed at static efficiency, attacks any monopoly position. As such, the process of business renewal may be hampered and progress inhibited. Furthermore, highly competitive environments may even lead to ruinous competition and 'a race to the bottom', leading to problems with quality, safety and compliance with the law. For the Dutch, striving to disentangle the distinction between collusion and collaboration has been an enlightening journey!

12 The trajectory of construction procurement in the UK

*Chris Goodier, Robby Soetanto, Andrew Fleming,
Peter McDermott and Simon Austin*

Summary

As highlighted in Chapter 5, procurement is a process and observable phenomenon entwined both culturally, politically and practically into the fabric and history of the construction industry. Historical reviews have highlighted recent changes in procurement systems that reflect the developments within the industry as a whole. This development is argued to be influenced mainly by a myriad of interconnected contextual drivers and issues both internal and external to the industry. The ability to adapt to change via a comprehensive understanding of these interconnected issues is a prerequisite for the industry to better meet the requirements of the society and deliver added value to the customers. This chapter gives initial research findings that identify procurement issues and trends from past construction reports, and uses these issues as a foundation on which to build future scenarios in the area of construction procurement, with particular reference to healthcare. Future scenarios constructed by experts in the area of construction procurement first depict two polarised paradigms, namely 'free market' and 'intervention', which will largely determine the future state of the industry. Then, second, the scenario of healthcare provision suggests a developmental shift from infrastructure production to service provision, and highlights the importance placed upon delivering end-user value.

Introduction

Procurement is commonly defined as the process of acquiring new products or services (Bower, 2003). There are many ways in which this process can be conducted, and it is influenced by a myriad of factors including social, political, technological and environmental. Understanding this process and the factors within their context is critical, due to the direct and indirect influence of procurement on the future competitiveness and well-being of the construction industry. These factors have caused changes in procurement practice in the past. In the late 1960s and 1970s, purchasing and procurement were generally considered to be a service to production (Farmer, 1997). Essentially, procurement was positioned in a supporting role to the manufacturing or production activities of the firm, an administrative, rather than a strategic activity

(Ansoff, 1970), and one which was purely transaction-based, quantified and specified. Market changes and events in the 1970s, such as the 1973 oil crisis and the impact of Japanese manufacturing, began to show the weaknesses of the traditionally disparate systems of purchase, supply and procurement. This blurring of roles continued and was mirrored in the organisational changes away from highly structured and rigid company structures to more multi-functional and cross-disciplinary companies. Indeed, the government and several industry reports had already identified the problems of relying only on contractual obligations in order to co-ordinate work (HMSO, 1964). Many regard that contractually based procurement practice has been largely responsible for the poor performance of the construction industry (e.g. Walker & Hampson, 2003).

In the mid-1990s procurement practices in construction began to undergo a considerable transformation, partly due to a shift in the business environments in which the procurement systems operate. In the early to mid 1990s, procurement experts and practitioners were involved mainly with debating the more strategic issues of the time. For example, these included privatisation, market liberalisation and the role of culture and trust in negotiations, as well as the more traditional themes of procurement systems, contractual arrangements and forms of contract (McDermott, 1999). In the late 1990s, some wider issues relating to procurement began to emerge, such as organisational learning and knowledge management, sustainable procurement and 'developmentally orientated procurement systems', i.e. those procurement systems/strategies that are charged with delivering wider social or economic benefits, rather than just cost and time criteria (McDermott, 2006). Recent reports also acknowledge that the 'softer' skills of persuasion and alignment are required by industry in order to best incorporate value creation and best practice in purchasing and procurement (FPA, 2003).

Today's clients and markets demand flexibility and innovation. Consequently, firms must be leaner, quicker and more proactive to keep their 'head above water' and in front of the competition. As a prime example, a holistic view of supply chain management is seen to provide the much needed competitive advantage. Procurement is no longer concentrating on operational activities, but on strategic objectives linked to the long-term survival and development of the organisations as a whole (Male, 2003). Using procurement as a 'competitive tool' brings many implications that have to be appropriately addressed before this tool can deliver the benefits promised. For instance, changing skill and attitudinal requirements permeate organisations within the supply chain. To this end, clients have to be more active and sophisticated, whereas supply chain members have to adopt collaborative working attitudes. Appropriate incentives and rewards have to be fairly decided and given when the common project objectives are achieved. This, in turn, will heighten the motivation and satisfaction of everybody involved, hence enhancing the likelihood of delivering added value for the project.

It seems that maximum benefits can only be achieved through a thorough understanding of all related issues/factors and their complex interconnectivities. This is the crux of an ongoing project called 'Sustained competitiveness in the UK construction sector: a fresh perspective', or the 'Big Ideas Project' for short, which is explained in the following section.

The 'Big Ideas': project and methodology

The Big Ideas Project[1] is a large multidisciplinary collaborative research project aimed at developing possible future development scenarios for the UK construction industry over the next 20 years, in order to support the industry in satisfying the future requirements of society. The central tenet is that a better understanding of the structure of underlying issues, events, barriers and trends through their causal relationships will enable the industry to address the persistent and deep-rooted problems that have hampered its performance for many decades. The overarching aim of the research is to develop multi-level strategic frameworks and policies for sustained competitiveness in the UK construction industry. This multi-level approach should address the full spectrum of firms within construction, including small subcontractors and suppliers that are often overlooked by the mainstream performance improvement agenda. The research is grounded on the real terrain in which these many firms operate through a thorough investigation of the current structural and cultural configurations of the industry.

The initial stages of this work involved reviewing the many construction futures reports which had been published between 1998 and 2005 in the UK and internationally (Harty et al., 2006a). More than 300 separate issues were identified from this literature and content analysis was used to group these in high-level clusters of related issues (Soetanto et al., 2006). These issues were used as a basis for developing future scenarios, which can be used as a tool to explore plausible states and pathways to an envisioned future. From this literature analysis and review, procurement was identified as one of a series of important issues for the future (Harty et al., 2006a). Interviews with two procurement experts were conducted to capture their perceived future scenarios in their chosen disciplines. These interviews, lasting about two hours, yielded detailed maps of issues, drivers and barriers, together with an associated recorded verbal narrative of the maps. This data was then converted into pictorial Visio™ maps (depicting relationships between issues, events, barriers and outcomes) and an associated textual explanation of the scenarios. Further detailed description of the methodology is presented elsewhere (Harty et al., 2006b).

The first interview scenario concerned two paradigms of 'free market' and 'intervention' and their impacts upon the possible developmental pathway of the industry. This resonated closely with the second interview regarding the trend in the procurement of healthcare facilities. On reflection, the first indicates the general trend in construction procurement, whereas the second provides a specific example of how the trend is enacted in practice. A brief historical development of this subject is presented below.

---

[1] The 'Sustained competitiveness in the UK construction sector: a fresh perspective', or the 'Big Ideas' for short, is a collaborative research project funded by the EPSRC and this support is gratefully acknowledged. The research team consists of Loughborough University, the University of Reading and the University of Salford. We are grateful to the input from these partners and to the experts that we interviewed as part of this chapter.

Trends in healthcare procurement

In May 2000, UK ministers launched *Sold on Health*, jointly with Her Majesty's Treasury and the Public Services Productivity Panel (NHS Estates, 2000). The document set out a range of programmes to improve the NHS's planning, procurement, operation and eventual disposal of its estate (NHS, 2006a). These include NHS ProCure21 as a direct response to the Government report *Achieving Excellence* (OGC, 2006). ProCure21 was developed by NHS Estates with the main objective of promoting better capital procurement in the NHS by developing a partnering programme using pre-accredited supply chains engaged in a long-term framework agreement. The aim was to cut out waste and duplication of effort in the tendering process, but also to bring the best of the construction industry together to deliver better value for money and in the end better clinical facilities for patients (NHS, 2006a). It was intended that ProCure21 would negate the need for traditional adversarial procurement and tendering by using pre-agreed supply chains and long-term framework agreements managed by principal supply chain partners (PSCPs). Under NHS ProCure21, it was recommended that the PSCPs should be involved in a project from the outset, to contribute to the planning and design phases, encouraging long-term collaborative working to achieve quality. In all, 230 projects are currently underway using ProCure21, with 54 completed (NHS, 2006b).

LIFT (Local Infrastructure Finance Trust) was announced in the NHS Plan in 2000 and involves private businesses taking over the ownership, financing and management of public sector infrastructure and services, and tying the public sector into exclusive long-term contracts with private sector companies (Unison, 2003). The plan was for health and social care premises (e.g. GPs' surgeries) to be built or refurbished and owned by new profit-making companies made up of public and private sector partners, with the private sector having a controlling interest. These would then be leased back to NHS bodies, GPs, local authorities, possibly voluntary sector and commercial organisations.

The Government's Sustainable Procurement Task Force was launched on 12 June 2006, charged with drawing up an action plan to bring about a step-change in sustainable public procurement so that the UK is among the leaders in the EU by 2009 (SPTF, 2006a). The Task Force recognised that this was important in moving the country towards a more sustainable economy. Partly, this is because the public sector's equivalent spend of 13% of GDP is capable of stimulating the market for more sustainable goods and services, and partly because only with government leadership can the consumption patterns of business and consumers be shifted onto a more sustainable path (SPTF, 2006b). The Task Force presented a National Action Plan with six key recommendations for the Government, namely: lead by example; set clear priorities; raise the bar; build capacity; remove barriers; and capture opportunities. The National Action Plan is intended to give the Government a clear direction on how to make real progress toward better, more sustainable procurement. In turn, this would allow it to move forward on sustainable development and set an example both to business and consumers in the UK and to other countries.

Organisations, such as Building Futures, have identified the future trends in the construction and healthcare industries in order to generate debate about their future paths. An example is the 2020 Vision research project, which identified the current social, economic and technological trends and how they might influence the design of healthcare environments over the next 20 years (Building Futures, 2001). They found that over the next 20 years the UK will experience:

- Very rapid developments in information and medical technology
- A demographic shift to an increasingly aged population
- Citizens becoming more informed about healthcare choices and decisions
- Modernisation of the health and construction industries (including new forms of procurement)
- New IT will change the location of different parts of the health service
- Public access to health information will continue to grow rapidly
- Tele-medicine will bring care closer to the patient

Other recent futures work has also looked at specific areas of the health sector, such as the 2029 report (IAF, 2005) which produced four alternative timelines for the future of biomedical R&D, together with recommendations for the future of healthcare provision.

A shift to service provision

More generally, many leading practitioners in UK construction are becoming aware of the increasing move away from product delivery towards the delivery of clients' needs through service provision. This shift from product delivery to service provision is already well established in aerospace, defence and manufacturing procurement. Several well-known traditional construction contractors have in recent years re-listed themselves on the Stock Exchange as service companies, the main reason being the changing procurement policies in both the public and private sectors. This, coupled with the extensive market segmentation in construction, means that the companies have to team up together in order to possess the necessary financial 'weight' to win these increasingly larger and larger contracts. Public–Private Partnership (PPP) and Private Finance Initiative (PFI) schemes have been extensively used for public sector procurement in the UK – such as hospitals, prisons and schools – and has resulted in a significant degree of supply-side consolidation and re-positioning of firms. 'Prime contracting' and other types of serial contracting arrangements provide the basis for clients to evaluate competence on the basis of service provision, the idea being that this approach encourages cost reductions by improving the capacity for supply-side innovation and increased efficiency. Concerns do remain however, regarding perceived value-for-money and design quality within PFI/PPP. This is mainly due to concerns on how much actual innovation and collaboration these procurement routes encourage. For it is suggested that much of the work within these consortia is still undertaken by disconnected teams and sub-contractors, and that the owner/operator/maintenance operations frequently find it difficult to influence the actual design decision-making process.

Hughes (2003) provides an insight into how the future of the industry might look if these trends in procurement and service provision continue:

> Integrated procurement systems became strategic alliances. Strategic alliances were formed in the name of partnership, mutual trust and collaborative working practices. Loosely based on limited networks of trading partners, they formed the basis for more formalized business relationships within groups of companies up and down the supply chain. Strategic alliances became mergers and acquisitions, increasing consolidation of the market into a few major conglomerates. These became so large that they were capable of funding PFI and PPP projects without the support of the banks, and selling completed schemes to pension funds provided them with the cash that they needed to invest in new ones. The consolidation of businesses affected the whole construction sector. As the trend toward leasing rather than buying gathered pace, most of the SMEs in the sector found that work dried up unless they joined in a strategic alliance. Eventually they were bought out or they simply went insolvent.

Two alternative future scenarios for construction healthcare procurement

Scenario 1: Two procurement paradigms and their possible outcomes (Figure 12.1)

The manner in which the procurement of services is planned and implemented has a great impact on the future state of society. Two contrasting scenarios are viewed to exist and are driven by the extent of control or governance. They are 'free market' and 'interventions accepted'. In many ways, they resemble two opposite paradigms of economic governance, the 'capitalist' and 'socialist'. In reality, the market situation (such as in the UK) is often in between these two extremes. The balance is influenced by many factors such as the political situation, leadership changes and the international money market. Possible future states are hypothesised as 'good' or 'bad' outcomes. The 'good' outcome is seen to be an industry that emphasises continuous performance improvement in cost, speed, quality, safety, sustainability, pattern of employment and community benefits. In contrast, the 'bad' outcome is an industry with poor time and cost performance, poor health and safety and with no consideration for sustainability (social, economic and environment). 'Free market' and 'interventions accepted' tend to, but do not necessarily yield, respectively, 'bad' and 'good' outcomes. A few exceptions occur however, that lead to the other way around. The two scenarios are described as follows.

Free market

The driver for the 'free market' trend emerged in 1968, when the Labour party introduced their monetary policy, under pressure from the International Monetary

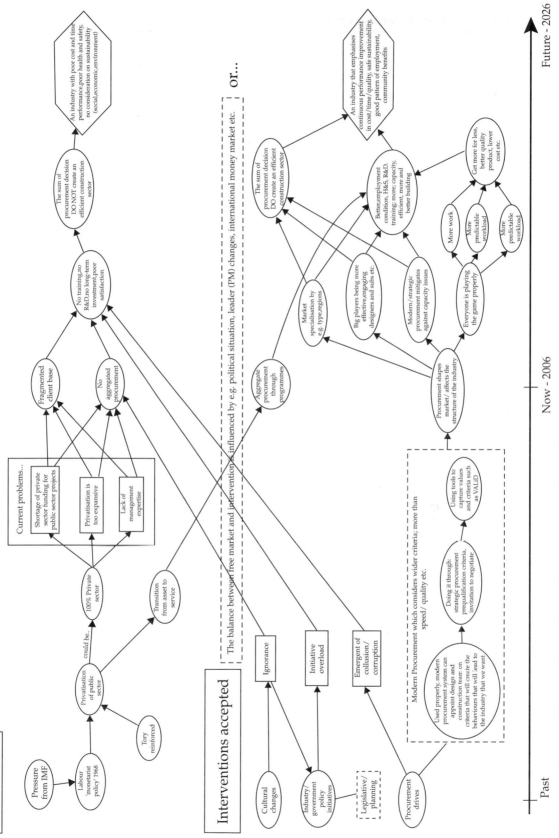

Figure 12.1 Two paradigms of 'free market' and 'intervention' and possible future outcomes.

Fund (IMF). This trend continued, reinforced by Thatcherism during the years of Conservative government. The principal reason for the privatisation of the public sector is the increasing demand for better quality infrastructure that can not be met by the public purse, and therefore needs a shared funding from private investment. To one extreme, this could lead to the private sector providing 100% of the required investment. Within this regime, several emerging problems are seen to exist, including shortage of funding, too expensive and lack of management expertise. These will exaggerate the level of fragmentation (especially the client base) and the absence of aggregate procurement. This free market may result in a lack of long-term investment including no training and R&D, and a poor level of satisfaction. Overall, the sum of procurement decisions does not create an efficient construction sector, leading to a 'bad' outcome. However, properly managed privatisation may yield to aggregate procurement, achieved through an integrated programme of deliveries, which may yield a 'good' outcome.

Acceptance of interventions

Interventions can have a positive and/or negative impact. On the negative side, industry/government policy initiatives (e.g. legislation/planning) can lead to 'initiative overload'. The difficulty of changing existing culture can negate the implementation of these initiatives and create an attitude of ignorance. Also, intervention-laden procurement may stimulate the emergence of collusion and corruption. All these will divert the intention of intervention towards a 'bad' outcome. A positive exemplar of intervention on procurement is modern procurement, which considers wider criteria, more than traditional cost, speed and quality. Used properly, modern procurement systems can appoint design and construction teams based upon criteria that will create the behaviours that will lead to the industry we desire. Modern procurement embraces principles of strategic procurement, pre-qualification criteria and invitation to negotiate, emphasising best value rather than lowest price competition. Recent research has commended the use of tools to capture stakeholders' values and criteria such as VALiD and DQI. This value-based procurement system will positively shape the market and the structure of the industry. Key players will concentrate on market specialisation by type (e.g. healthcare) and region. They will be more effective and efficient, and engage designers, suppliers and subcontractors in a real sense in their quest for best value. Modern/strategic procurement will militate against capacity issues where everyone is 'playing the game properly', leading to more work and a more predictable workload. This will deliver more for less and a better quality product at lower cost. The resultant outcome will be a superior industry, in terms of efficiency, employment, H&S, R&D and more and better buildings for society.

Scenario 2: Trend of commissioning services (Figure 12.2)

The biggest driver in the healthcare business at the moment is the procurement of services, not facilities, which has an enormous impact on the supply chain. This trend is towards output rather than specification. For some, this is synonymous with 'backdoor privatisation' as if privatisation in the Thatcher era has emerged in a different guise. For

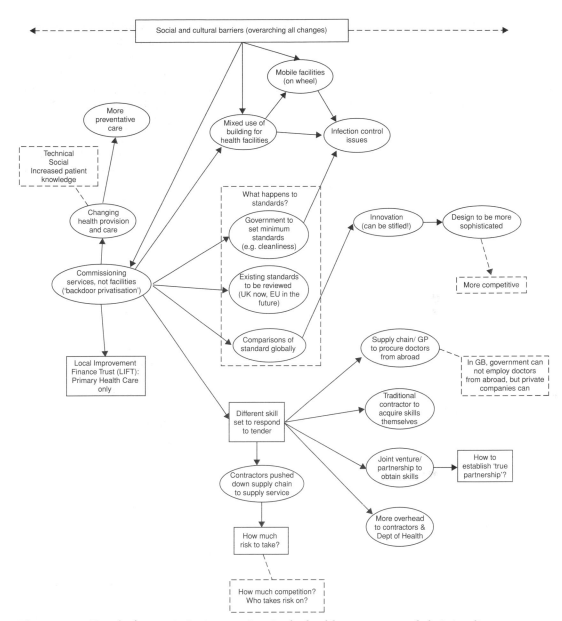

Figure 12.2 Trend of commissioning services in the healthcare sector and their implications.

the 'contracting' business, this trend requires different skill sets to respond to tender. Contractors are being 'pushed down' the supply chain to service suppliers. This would also mean that they might wish or need to establish partnerships and/or joint ventures to obtain the necessary skills. For example, healthcare providers may wish to procure the services of doctors from abroad (e.g. Poland). However, by law, EU governments cannot employ doctors from other countries, but private sectors are permitted to do so.

Several potential barriers should also be considered by the potential contractors, such as establishing 'true partnership', the amount of risk to take, the allocation of risk to appropriate parties and level of competition.

The pertinent question concerning this major shift is the quality of service provided, i.e. what happens to the standards? Government will set minimum standards for services and review the existing standards in the UK. In the future, comparison of standards will be undertaken against EU countries and others globally. These comparisons will stimulate innovations and enhance design, as well as competition. This trend opens up alternatives for the provision of construction-related services, such as the mixed use of buildings and the use of mobile facilities. Although these alternatives may provide flexibility and be an economical solution, issues such as infection control do exist.

Overall, the trend towards providing services will dramatically change both health provision and care, but there are technical, social and cultural barriers that need to be overcome.

Future modelling of procurement

So, the scenarios will provide an insight into the construction industry's behaviour to enable key policy makers to review existing policies and determine appropriate policies for future implementation. Systems dynamics (SD) will then be used to build on this and help interpret the findings from the classical literature and construction company case studies that have been undertaken as part of the 'Big Ideas' project, the initial results of which are published elsewhere (Fleming et al., 2006).

Conclusions

This chapter has presented some background on the history and present status of construction procurement in the UK healthcare sector. Two future scenarios devised by experts in the area of construction procurement have been presented which depict a general shift from infrastructure production to service provision, with one providing examples of two possible future markets: 'free market' and 'intervention'.

The purpose of these scenarios is not to predict which is going to happen, but to generate open debate amongst practitioners and policy makers so that plans can be put in place to ensure that the industry progresses towards the more 'desirable' future, whilst simultaneously being aware of the risks associated with the 'less-desirable' scenario. The scenarios are considered as stakeholder sensitive. When discussing possible future scenarios, it should also be remembered that one individual's 'desirable' outcome can often be another's 'undesirable'.

These considerations begin to illuminate the systemic impacts of trends in procurement. In order to understand the internal interconnectivities within the industry and the external interconnectivities between the industry and its environment and how these interrelate, an SD methodology will be employed to enable systemic interconnectivities to be modelled and analysed.

13 Delivering full value through seamless information systems

Ghassan Aouad, Nick Bakis, Song Wu and Emeka Efe Osaji

Summary

The recurring theme in the construction sector, that of closer collaboration and integration, was introduced in Chapter 6. The present chapter examines the nD modelling platform, developed at the University of Salford, that aims to promote such closer collaboration and integration in the design process. The nD modelling platform integrates the various aspects (dimensions) of design into a single, holistic model, enabling the different design professionals to work closely together in developing high-value solutions through effective experimentation with ideas. A significant part of the chapter is devoted to examining the business benefits of nD modelling – a key implementation issue as shown in Chapter 11. Recognising the fact that the business benefits of any information system in general cannot usually be expressed in absolute terms, and considering why this is the case, the chapter provides a comprehensive list of the potential benefits of nD modelling. The assessment of the business value of this technology is, however, left to each particular organisation, although some remarks in relation to this assessment are made.

Introduction

The use of IT in the construction industry is characterised by a high degree of fragmentation, reflecting the complex, differentiated nature of the industry itself. Most of the IT applications in construction today are autonomous as they have been developed by different vendors to support particular tasks in isolation. They operate independently, use their own media and formats to store information, and have very poor capabilities in utilising each other's data directly in digital form, even though large parts of those data are closely interrelated as they refer to different parts or aspects of the same building. As a result, human intervention is required to transfer information from one application to the other. In addition, as the information in each application is kept independently, care must be taken in order to maintain information consistency.

Since the mid-1980s, a substantial body of research has been aimed at bridging such 'islands of automation' and bringing closer an integrated use of IT in the construction industry (Bakis et al., 2005). An integrated use of IT should improve efficiency and minimise communication errors. However, the biggest benefits of integration will come not mainly by assisting and speeding up existing processes, but rather by facilitating a more integrated project process that brings together the full design and construction team as envisaged by Egan (1998). Only through such a process can waste be significantly reduced and both quality and efficiency improved in order to deliver value (Egan, 1998).

In the design process, an integrated use of IT can facilitate integration by minimising the costs, delays and errors associated with the manual management and exchange of information, thus providing the opportunity for a more iterative and collaborative design effort. In such an effort, the design team works closely together to gradually refine the design by systematically examining all its aspects and experimenting with ideas. While, however, an integrated information management and exchange can facilitate a more systematic 'what-if' analysis during the design effort, the results of this analysis may be sub-optimal, as the applications still operate independently and improvements in one area may result in worsening in another. In order to enable a true 'what-if' analysis where the design is optimised as a whole, an additional integration layer is required to bring and enable examination of the various aspects of the design together.

The nD modelling platform, developed at the University of Salford, aims to provide such an additional layer by integrating the various aspects (dimensions) of design into a single holistic model that portrays and enables the examination of the design as a whole over its complete life-cycle. This chapter examines the nD modelling platform and the opportunities it provides in improving the performance of construction organisations. In examining these opportunities, the chapter recognises that how these opportunities will materialise depends on many different factors. As explained below, the business value of an information system is a matter of context and perception and rather difficult to determine or express in absolute terms. Each organisation must carefully examine how the business value of any information technology in general and nD modelling in particular applies to its particular situation, and so the emphasis is on how this evaluation might take place.

This chapter consists of two main parts. The first part examines the nD modelling platform, focusing on its capabilities and development as well as the main drivers and barriers to its application in the construction industry. The second part examines the opportunities offered by nD modelling in improving the performance of construction organisations. In particular, it first considers why the business value of information systems cannot be easily determined or expressed in absolute terms. It then considers the potential benefits of nD modelling, leaving the assessment of the business value of this technology to each particular organisation.

The nD modelling platform

The terms 'nD modelling' and 'nD CAD' are gaining increased usage in the field of information communication technologies (ICTs) and building design; the concept gaining a heightened profile via the £0.5 million EPSRC-funded 3D to nD modelling project at the University of Salford. An nD model is an extension of the building information model that incorporates all the design information required at each stage of the life-cycle of a building facility (Lee et al., 2003). Thus, a building information model (BIM) is a computer model database of building design information, which may also contain information about the building's construction, management, operations and maintenance (Graphisoft, 2003). This database allows different views of the information to be generated automatically, views that correspond to traditional design documents such as plans, sections, elevations and schedules. As the documents are derived from the same database, they are all co-ordinated and accurate – any design changes made in the model will automatically be reflected in the resulting drawings, ensuring a complete and consistent set of documentation (Graphisoft, 2003).

Both 2D and 3D modelling in the construction industry take their provenance from the laws governing the positioning and dimensions of a point or object in physics, whereby: a three-number vector represents a point in space; the x and y axes describe the planar state; and the z axis depicts the height (Lee et al., 2003). Further, 3D modelling in construction goes beyond the object's geometric dimensions and replicates visual attributes such as colour and texture. This visualisation is a common attribute of many AEC design packages, such as 3D Studio Max and ArchiCAD, which enable the simulation of reality in all its aspects or allow a rehearsal medium for strategic planning. Combining time sequencing in visual environments with the 3D geometric model (x, y, z) is commonly referred to as '4D CAD/modelling' (Rischmoller et al., 2000). Using 4D CAD, the processes of building construction can be demonstrated before any real construction activities occur (Kunz et al., 2002). This will help users to find the possible mistakes and conflicts at the early stage of a construction project, and to enable stakeholders to predict the construction schedule. Research projects around the world have taken the concept and developed it further, so that software prototypes and commercial packages have begun to emerge. In the USA, the Centre of Integrated Facility Engineering (CIFE) at Stanford University has implemented the concept of the 4D model on the Walt Disney Concert Hall project. In the UK, the University of Teesside's VIRCON project integrates a comprehensive core database designed with standard classification methods (Uniclass) with a CAD package, a project management package (MS Project) and graphical user interfaces as a 4D/VR model. This has been used to simulate the construction processes of an £8 million, three-storey development for the University's Health School (Dawood et al., 2002). Commercial packages are also now available, such as 4D Simulation from VirtualStep, Schedule Simulator from Bentley and the 4D CAD System from the JGC Corporation.

nD modelling builds on the concept of 4D modelling by integrating '*n*' number of design dimensions into a holistic model, which would enable users to portray and visually project the building design over its complete life-cycle. nD modelling is based upon the building information model (BIM), a concept first introduced in the 1970s and the basis of considerable research in construction IT ever since. The idea evolved with the introduction of object-oriented CAD; the 'objects' in these CAD systems (e.g. doors, walls, windows, roofs) can also store non-graphical data about the building in a logical structure. The BIM is a repository that stores all the data 'objects', with each object being described only once. Both graphical and non-graphical documents, such as drawings and specifications, schedules and other data, respectively, are included.

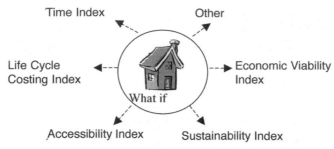

Figure 13.1   'What-if' analysis indexes of the 3D to nD modelling project.

Changes to each item are made in only one place and so each project participant sees the same information in the repository. By handling project documentation in this way, communication problems that slow down projects and increase costs can be greatly reduced (Cyon Research, 2003). Leading CAD vendors such as AutoDesk, Bentley and Graphisoft have promoted BIM heavily with their own BIM solutions and demonstrated the benefits of the concept. However, as these solutions are based on different, non-compatible standards, an open and neutral data format is required to ensure data compatibility across the different applications that project participants may be using.

Industry Foundation Classes (IFC), developed by the International Alliance for Interoperability (IAI), provide such capabilities. IFCs provide a set of rules and protocols that determine how the data representing the building in the model are defined, and the agreed specification of classes of components enables the development of a common language for construction. IFC-based objects allow project models to be shared whilst allowing each profession to define its own view of the objects contained in that model. This leads to improved efficiency in cost estimating, building services design, construction and facility management. IFCs enable interoperability between the various AEC/FM software applications, allowing software developers to use IFCs to create applications that use universal objects based on the IFC specification. Furthermore, this shared data can continue to evolve after the design phase and throughout the construction and occupation of the building.

3D to nD modelling project

The 3D to nD research project, at the University of Salford, has developed a multi-dimensional computer model that will portray and visually project the entire design and construction process, enabling users to 'see' and simulate the whole life of the project. This, it is anticipated, will help to improve decision-making processes and construction performance by enabling true 'what-if' analysis to be performed to demonstrate the real cost in terms of the variables of the design issues (Figure 13.1). Therefore, the trade-offs between the parameters can be clearly envisaged:

- Predict and plan the construction process
- Determine cost options
- Maximise sustainability

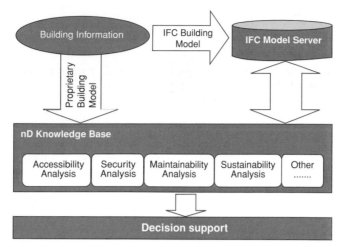

Figure 13.2   System architecture of the nD modelling prototype tool.

- Investigate energy requirements
- Examine people's accessibility
- Determine maintenance needs
- Incorporate crime deterrent features
- Examine the building's acoustics

The project aimed to develop the infrastructure, methodologies and technologies to facilitate the integration of time, cost, accessibility, sustainability, maintainability, acoustics, crime deterrent and thermal requirements. It assembled and combined the leading advances that had been made in discrete information communication technologies (ICTs) and process improvement to produce an integrated prototyping platform for the construction and engineering industries. This output will allow seamless communication, simulation and visualisation, and intelligent and dynamic interaction of emerging building design prototypes, so that their fitness of purpose for economic, environmental, building performance and human usability can be considered in an integrated manner. Conceptually, this involves taking 3D modelling in the built environment to an *n*th number of dimensions.

The developed nD tool's system architecture is illustrated in Figure 13.2.

nD knowledge base

This is a platform that provides information analysis services for the design knowledge related to the various design perspective constraints of the nD modelling (i.e. accessibility requirements, crime deterrent measures, sustainability requirements, etc.). Information from various design handbooks and guidelines on the legislative specifications of building components will be used, together with physical building data from a building information model to perform individual analysis.

Decision support

The support for the decision-making process has proved to be problematic. Traditionally, a whole host of construction specialists are involved in instigating the design of modern buildings. With so much information coming from so many experts, it becomes very difficult for the client to visualise the design, the changes applied and the subsequent impacts on the time and cost of the construction project. Changing and adapting design and planning schedules and cost estimates to aid client decision-making can be laborious, time-consuming and costly. Each of the design parameters that the stakeholders seek to consider will have a host of social, economic and legislative constraints that may be in conflict with one another. Furthermore, as each of these factors vary – in the amount and type of impacts they can have – they will have a direct impact on the time and cost of the construction project.

The criteria for successful design, therefore, will include a measure of the extent to which all these factors can be co-ordinated and mutually satisfied to meet the expectations of all the parties involved. Multi-criterion decision analysis (MCDA) techniques have been adopted to tackle the problem. The solution for the combined assessment of qualitative criteria (i.e. criteria from the Building Regulations and British Standard documents that cannot be directly measured against in their present form) and quantitative criteria (e.g. expressed in geometric dimensions, monetary units, etc.) has also been investigated. Analytic Hierarchy Process (AHP) has been used to assess both qualitative criteria (i.e. criteria that cannot be directly measured) and quantitative criteria (e.g. expressed in dimensions, monetary units, etc.). However, the common understanding and definition of the similar concepts in the construction project from different specialists is yet to be developed, and is required to make the decision-making process more interactive and intelligent.

So far, the nD prototype tool incorporates whole-life-cycle costing (using data generated by Salford's Life-Cycle Costing Research Project), acoustics (using the *Rw* weighted sound reduction index), environmental impact data (using BRE's *Green Guide to Specification Data*), crime (using the Secured by Design Scheme standards) and accessibility (using BSI:83001). Technology for space analysis has been developed to support the accessibility analysis and techniques for determining suitable building material in accordance with life-cycle costing, acoustics and environmental impact factors.

Barriers and drivers for the development and implementation of nD modelling

The barriers and drivers for the development and implementation of nD modelling have been discussed by a large number of international participants in nD research workshops (Lee et al., 2005). In general, the participants have all agreed that there are no specifically regional or national barriers related directly to the uptake of nD modelling. That is, the challenge of nD modelling is global and the barriers and enablers quite generic. However, the ease and speed of implementation varies depending on the level of involvement from both the construction and IT industries. Thus, a greater need for collaboration is paramount. Key barriers and drivers are summarised below.

Lack of collaboration

It was noted that there were various overlaps in the research and development being undertaken across the world. The need for more collaboration was seen as a burning issue among the research groups and that resolving this was a crucial priority. It was stated that for true benefits to be realised, everybody concerned needs to get involved (Lee et al., 2003). However, it was thought that this would bring additional problems. For this inclusive approach would mean incremental adoption of the relevant technologies, which would create a 'chicken and egg' situation: there would need to be a critical mass of existing technology users to encourage others and to reap the full benefits of nD modelling.

Readiness of the technology

Both the readiness of the technology, based on its current usage rates, and the amount of development that was needed to make it ready for widespread use have been questioned. Our research suggested that, although there is investment to overcome this barrier, the software is not currently readily prevalent in the market. Part of the reason for this is the need for the concept of nD to be transformed into an actual product. An additional means of overcoming this barrier would be financial initiatives. The development of industry–university relationships was cited as a key requirement, but the way that academia currently thinks about industry is providing an additional cultural barrier for this relationship. However, it was not thought that these relationship barriers are insurmountable: the belief was expressed that industry is ready to change, and that improved education programmes would provide the mechanism for this. However, this would require the involvement of the education providers and, perhaps more fundamentally, the industry managers who would have a role preparing the workforce in terms of exposure to the new ways of working and thinking that nD and object modelling requires. However, even with the best of intentions, changes to education and training and the effects of these might take time to run their natural course in circumstances where advocates do not have as much control as they might like.

Lack of education

Whilst lack of education itself is currently a barrier, it was argued that it is as important to make explicit exactly what the knowledge is that needs to be passed on. There is a lack of more general understanding of how things work in the present, never mind how they should work in an idealised future. More specifically, there is a lack of understanding of the practicalities of working in multi-disciplinary groups and the kinds of processes that will be both necessary to facilitate this and become a resultant outcome of such working. In a similar vein, there was a cited lack of understanding of the nD process and data management that would be required through the model's life-cycle, and even a lack of understanding of the data required at each stage of the process for the necessary decision-making processes to occur. The reason for this was given as being due to the heavy document focus of the industry at this time. All these barriers emphasise the importance of understanding nD and its implications for construction as a new set of processes first and foremost, before these new processes are mediated and supported by technology.

Over focus on technology

Like the importance ascribed above to understanding the processes, data needs and management issues before the automation of the processes in a 'what-if' model, the emphasis should be upon understanding the skills needed to assimilate nD into current working practices without undue technology push. It was thought that there is too great a focus upon central databases and that this is in contrast to the amount of development actually taking place on data and data standards such as STEP and IFCs, which are being adopted.

Lack of performance data

It is generally agreed that the lack of performance data that currently exists to demonstrate value, or lack of this, is itself a consequence of the lack of nD process. The lack of large, integrating projects demonstrating the potential opportunities was also cited as a barrier, as was the lack of incentive programmes more generally. The industry needs a greater public demonstration of the economic benefits that nD processes and technologies could enable. The research suggested that there might be a communication problem, and this might be adding to the confusion over the need for change based on the way it is framing the problem. (nD and, more specifically, nD technologies have been framed as a solution to a specific industry recognised problem, not as one of numerous solutions for its more generic problems.) This situation is probably exacerbated by the cited lack of a dominant client, as compared to the aircraft industry. The use of the term '4D modelling' was also deemed to be too academic and that comprehension would be aided by talking about scheduling instead.

Legal implications

The legal implications of the incorrect use of data in the public domain are a major concern. Not only can it have effects on an individual basis – problems of protection of professional boundaries – but worries about such problems then impact upon the organisation more generally and can detrimentally affect culture change, cooperation, effectiveness of team-working and so on.

Enablers and drivers

New forms of contract

In terms of the enablers for the development and implementation of nD modelling, there was agreement upon the importance of new forms of contract as a driver. The move from traditional to new contract forms – such as PFIs and PPPs – puts the contractor more in the role of a service provider, and this is differentiated from the role of the constructor. These new forms of contract were cited as providing drivers to exploit the available technology and other vehicles for integrated teams. This need is becoming ever more necessary as changing procurement processes are forcing people used to working in sequential ways to collaborate with one another. The new contract forms prevalent within the UK include PPPs, PFIs, LIFT (which is service, procurement and contractual arrangements for healthcare) and LEP (Local Education Partnership).

Change in approach

The other issue on which there is agreement, is the need for a change in approach. A goal-based approach is needed that understands individuals' information require-ments for performing their roles. This is dependent upon being able to understand the roles that people play in their (integrated) teams. A team-based approach should go further than fostering and encouraging collaboration and examining roles, and should enable the individual to perform a role in any part of the life-cycle of the construction process. For this to be achieved, the correlations and relationships between roles and processes need to be understood.

Industry champions

In terms of new approaches, a move towards enterprise agendas was also suggested including support by industry champions, whose role it would be to convince individ-uals of the importance of future visions and pathways for the industry of the future. It is essential to establish the route that needs to be followed in order to reach a vision of the construction industry of the future, and that this needs to be the emphasis over the current approach of just selling the endpoint. As well as drivers from within the industry, the Government was also cited as an institution that could and should influ-ence change from outside the industry by setting standards and developing enabling infrastructures.

nD modelling and opportunities for improved performance

Before examining the opportunities that nD modelling provides to improving the per-formance of construction organisations, it is important to recognise that how these opportunities will materialise depends on many different factors.

The elusive nature of IT business value

As Remenyi (2000) notes, the assessment of the business value of information systems can become elusive. Not all the business benefits of an information system might ma-terialise and new, unexpected benefits might appear during the system's lifetime. As several authors note (Farbey et al., 1993; Mooney et al., 1995; Soh & Markus, 1995; Remenyi et al., 2000), an information system in general has no direct value in its own right. Instead, an information system has a potential for derived value. This derived value is directly linked to the business objectives and processes that the system aims to support. For instance, the value of a system that improves customer service is directly linked to the value that an organisation places on this improved customer service. Or, as another example, nD modelling aims to improve the collaboration between the participants in a construction project. However, with the traditional forms of procurement and sequential design and construction process, this improvement will be kept to minimum. Only within an environment that promotes collaboration and the open and free exchange of information, such as partnering (Bennett & Jayes, 1998), can the full potential of nD modelling be realised.

The fact that an information system has no direct value in its own right means that the business value of an information system in general and nD modelling in particular cannot be perceived directly or on its own, independently of the business objectives and processes the system aims to support. It is those objectives and processes that would primarily determine the contribution of the system to the organisation's performance. The fact that an information system has no direct value in its own right also means that we cannot judge the success of an information system solely based on the improvement of the organisation's performance. As Soh & Markus (1995) note, the success of an information system should be ultimately judged on how well the system supports a specific business initiative and not on how well this initiative performs. In the above example, for instance, the information system should be judged on whether higher levels of customer service have been achieved and not on whether those higher levels of customer service have led to increased revenue. Before investing in nD modelling, therefore, construction organisations need to consider their investment within a broader business practice and process improvement effort. Within the context of such an effort, nD modelling becomes the enabler rather than the creator of business value.

The role of IT since the 1990s has dramatically shifted, from mainly automating clerical tasks (efficiency zone) to performing tasks more effectively or even changing the way organisations conduct business (effectiveness and business transformation zones). However, several of the business benefits of an information system might be intangible, in the sense that they do not directly lead to identifiable performance improvements. For instance, an information system might improve customer satisfaction, which is difficult to translate into financial or any other measurable terms. As intangible benefits are difficult to quantify, the business value of modern information systems is more of a subjective than objective nature.

Of course, an organisation might opt for an 'objective' but partial assessment of the business value of an information system, in strict financial terms, by considering only those benefits that can be directly attributed to cost savings or the generation of additional revenue. Several examples of such organisations are reported in the literature (Ballantine et al., 1994; Ballantine & Stray, 1999). Using standard cost-accounting techniques, such as 'net present value' (NPV), 'discounted cash flow' (DCF), or 'payback period' (Ballantine et al., 1994), those organisations determine the worthiness of the investment by comparing the cost savings or additional revenue that can be attributed to the system with the costs of developing, maintaining and operating the system.

However, as Farbey et al. (1993) note, any organisation that claims to apply IT strategically but only considers quantifiable benefits is out of alignment. Accounting techniques and strict financial measures might be appropriate for systems that mainly automate clerical tasks, but as the role of IT has shifted from the efficiency to the effectiveness and business transformation zones, there is a need for a broader view of IT business value.

As there could be several stakeholders affected by the introduction of an information system, IT value usually has multiple perspectives. These perspectives could be quite diverse, as the different stakeholders may have different and often conflicting objectives. For example, in a case study reported by Smithson & Hirschheim (1998), the managers of an organisation wanted to outsource the IT function in order to reduce costs. The end-users, on the other hand, wanted to maintain the existing IT department, which was providing a high-quality service. At the end, the decision to outsource the IT function was taken, and while the managers saw this decision as successful, the users were highly disappointed by the quality of service they were receiving from the outsourcing organisation. The different perspectives of the various stakeholders must be taken into account when evaluating an information system. All key stakeholders must be identified and, if possible, consulted about their view of the system's value. Without engaging all key stakeholders, there is a danger that the stakeholders will oppose the introduction or obstruct the operation of the system, limiting its potential benefits. Of course, the objectives of the various stakeholders may always remain conflicting and a compromise may not always be possible. Nevertheless, identifying any problems in advance is particularly important before committing to any significant investment.

The business value of information systems does not usually stay static, but rather evolves and is dynamic. As Remenyi (2000) points out, an information system may provide the basis for additional functionality or it may be used in ways that have not been previously thought of. This is particularly true for any infrastructure investments (such as nD modelling), which do not deliver any benefits directly but provide the basis for the operation of other applications. As the role of the system within the organisation is gradually understood, refinements could be made and new benefits arise. However, as Remenyi (2000) adds, trying to forecast those benefits is almost impossible, especially in today's dynamic economic environment where businesses rapidly change.

The introduction of a substantial information system may also bring significant changes in organisational terms (e.g. changes in organisational structure or procedures), social terms (e.g. social interaction, quality of working life, organisational culture) and management terms (e.g. information access and decision-making) (Walsham, 1993). These changes might happen gradually and take a long time to materialise, and might affect the organisation's performance in a positive or negative way. Trying to envisage those changes and the benefits they might bring is a challenging task. But even if those benefits could be identified, it is rather difficult, as Farbey et al. (1993) note, to isolate the factors that contribute to them and establish whether they should be attributed to the information system or not.

Even if all the potential benefits of an information system can be identified, one should be aware that not all of these benefits might materialise. For example, as Brynjolfsson (1993) points out, an information system might not materialise all of its benefits simply because it is not used properly. For instance, learning difficulties may cause a lag in delivering the benefits. Or, as noted above, stakeholders that oppose the system's introduction, such as users feeling their jobs are at risk, may obstruct the system's operation, thus limiting its benefits.

Potential benefits of nD modelling

The fact that the business value of an information system depends on the business objectives and processes that the system aims to support, as well as the fact that the business value of an information system is a matter of perception, means that only the potential benefits of an information system can be indicated. Each organisation should assess how these benefits apply to its own particular situation. In addition, due to the difficulties in identifying the business benefits of an information system (e.g. due to the evolving nature of information systems or the fact that not all benefits might materialise), any list of the potential benefits should not be considered as definite or exhaustive. Instead, it should be considered as a starting point in identifying the benefits and assisting the evaluation effort.

Construct IT distinguishes the benefits into efficiency (can be measured in monetary terms), effectiveness (can be assigned a subjective value) and performance (strategic – no value assigned). For each of those categories, it lists the benefits per business process to which the benefits apply. The benefits that might be used from Construct-IT (1998) are shown below.

Efficiency benefits

- **Marketing:** Reduced marketing costs, ability to handle more enquiries
- **Information management:** Reduced communication costs, reduced paperwork
- **Client management:** Quicker response to client enquiries
- **Design:** Reduced lead times for design, reduced rework, increased information exchange
- **Construction:** Reduced construction times, improved productivity, reduced waste (by more systematically examining constructability issues).

Effectiveness benefits

- **Business planning:** Increased sales, minimisation of business risk, strategic competitive advantage, maintenance of competitive capacity
- **Marketing:** Improved company image, generation of new business
- **Information management:** Fewer information bottlenecks
- **Finance:** Minimisation of business risk
- **Client management:** Improved quality of output, faster delivery of services, improved focus on client requirements
- **Design:** Improved quality of output, increased speed of new design development, faster response to design changes
- **Operation and maintenance:** Ability to refer back to data

Performance benefits

- **Client management:** Improved information exchange with clients, increased client satisfaction, strategic competitive advantage
- **Design:** Improved data sharing among project teams
- **Operation and management:** Improved capture of design decisions

Assessing the business value of nD modelling

While the above benefits provide an indication of the potential value of nD modelling, the business value of this technology to any particular organisation should be carefully examined. Ideally, a comprehensive evaluation will be carried out. This evaluation should ideally be performed by conducting a pilot study involving the use of the system, or, if this is not possible, by carefully examining the system's benefits, costs and risks. Ideally, all key stakeholders affected by the introduction of the system should be involved in the evaluation exercise.

Two main approaches could be followed in the evaluation exercise: one that is conducted at a coarse, project-level or organisation-level of detail, and one that is conducted at a more analytical, process-level of detail. With the evaluation that is conducted at a project-level or organisation-level of detail, the impact of the system to a project or organisation as a whole is considered. For instance, if a pilot study involving the use of the system can be conducted then project-level output measures could be used – such as the time it takes to complete a project – in order to compare a set of projects that are completed with the assistance of the system with a similar set of projects that are completed without. Similarly, a survey could be conducted to examine the impact of a system on a number of organisations as a whole (for example, how the organisations' profitability has been affected by the introduction of the system). The problem, however, with such an approach is that, as several authors note (McKeen et al., 1999; Tallon et al., 2001), it is not clear whether any difference in performance is owed to the system or any other factors. In addition, as Mooney et al. (1995) note, such an approach provides no insight on how and why the business value is created.

An alternative approach to evaluation, advocated by authors such as Mukhopadhyay & Cooper, (1993), Mende et al. (1994), Mooney et al. (1995), McKeen et al. (1999) and Tallon et al. (2001), is to take a more analytical approach and try to understand how the system creates business value by examining the impact of the system on each individual business process. For example, how does the system improve the procurement process, the design process, the construction process or any other process, at this or any other level of detail? By establishing the impact of the system on each individual process, detailed measures of the system's performance at the level of each process could be derived. By aggregating the benefits from each individual process, an overall picture of the system's value could be obtained. If the evaluation is undertaken within the context of a process improvement or re-engineering effort, the *as-is* and *to-be* process models could be used to guide the analysis. By comparing the two models, the impact of the system can be examined in a systematic manner.

Such a process-oriented approach to evaluation, that analyses the impact of the system at the level where the system is deployed, provides a greater insight on how and why business value is created. As Mooney et al. (1995) note, such an approach 'enables [one] to move beyond correlation evidence to explanation of the technological features, process characteristics and organisational settings conducive to producing IT business value'.

If we examine the list of potential benefits of nD modelling presented above, we will see that several of those benefits are intangible, in the sense that they do not directly lead to identifiable performance improvements. So, if the business value of nD modelling and any information technology in general cannot be expressed in single monetary terms, how could the worthiness of the investment be decided? As we saw, several companies consider only the financial benefits of the system but, as several authors note, intangible benefits are too important to ignore.

When comparing two alternative systems, a numeric indication of the system's value could be obtained by assigning a subjective value to each intangible benefit. For example, a weight could be assigned to denote the importance of each of the system's potential effects, e.g. 'improved work conditions', and a rating used to denote the system's potential impact on each of those effects, 'e.g. the work conditions would be probably improved by 70%'. By multiplying each weight with the corresponding rating, a score could be derived for each benefit, and by adding all the scores, a numeric indication of the system's total impact could be obtained (Parker & Benson, 1998; ConstructIT, 1998(b)). However, as Svavarsson et al. (2002) note, while such an approach could be useful in comparing alternative systems, it is of little help in deciding the worthiness of a particular investment, since the calculated score provides no indication of the system's true value.

So without an objective measure of the system's value, the question of how to evaluate an IT investment still remains. However, the absence of a single objective measure should not necessarily create a problem. As several authors argue, it is preferable to draw a more 'balanced' and overall picture of the system's value by presenting several different criteria and let the decision-maker judge the worthiness of the investment in an instinctive and intuitive manner. For example, Willcocks & Lester (1994) propose the use of the 'balanced scorecard' approach of Kaplan and Norton (1992) to examine the contribution of the information system from four different perspectives: the 'financial perspective', the 'internal business perspective', the 'customer perspective' and the 'innovation and learning perspective'. The benefits from each perspective are listed separately, with no attempt to aggregate them. In this way, a more 'balanced' view of the contribution of the system can be provided and the decision on the worthiness of the investment left to the decision-maker. Similarly, Construct IT (1998(b)), in its *Measuring the Benefits of IT Innovation* framework, attempts no quantification of strategic benefits; rather, it lists those benefits and lets the decision-maker judge the worthiness of the investment.

As Bannister & Remenyi (1999) point out, such an approach to decision-making, based on instinct and intuition, might sound disturbing to someone who is a formalist. However, as they argue, there is no fundamental reason why this should be the case. As they point out:

instinct after all is a central part of many decision making processes and especially the management decision making process ... It is not something to be condemned but rather a different and more subtle kind of reasoning – a method of taking into account how the world really is rather than simply what the spreadsheets say ... It is something to be celebrated as part of not only that which differentiates man from machine but separates mediocre from top flight management.

While some organisations might agree with the above comments and accept a subjective approach to evaluation, other organisations might lean towards the more objective end. At the objective end, fewer risks are taken but greater opportunities might be missed. At the subjective end, the opposite holds. With each approach having its own merit, the decision of which approach to follow rests to each particular organisation.

Conclusions

A recurring theme is that of closer collaboration and integration through which greater value can be achieved. This chapter has examined the nD modelling platform, which aims to promote such closer collaboration and integration in the design process. The nD modelling platform provides the basis for integrating the various aspects of design into a single holistic model, enabling the design professionals to work closer together in developing high-value solutions through effective experimentation with ideas.

As we have seen in this chapter, in addition to supporting a more collaborative and integrated design effort, nD modelling provides several other opportunities for improving the performance of construction organisations. However, as the business value of any information system depends on the business objectives and processes that the system aims to support, a careful evaluation must be carried out before investing in nD modelling. Two main approaches that could be followed in such an evaluation have been set out: one at the project or organisation level, and one at the process level. A process level approach to evaluation provides a greater insight into how and why business value is created. In assessing the business value of nD modelling and any information technology in general, an objective approach to evaluation might be followed, trying to establish a monetary value. Another organisation might choose to follow a subjective approach, trusting human judgement.

14 Long-term educational implications of revaluing construction

Melvyn A. Lees

Introduction, background and historical development

This chapter builds on Chapter 7 by examining the long-term implications for education that will result from the revaluing construction agenda. It takes higher education in the field of construction and construction management as its focus but, where appropriate, the picture is widened to encompass all disciplines with a connection to the built environment.

The chapter starts, in this section, with a look at the direction in which construction education has been moving since the 1960s. This is followed with a section that looks at how well construction education is performing. The next section then exposes the traditional dichotomy between education and training. In the final section, the need for greater engagement between industry and academia is argued for as the basis for improving the value of education. Finally, some conclusions are drawn from the debate and offered for action.

What is the purpose of education? Martin Luther King (1948) wrote:

> It seems to me that education has a two-fold function to perform in the life of man and in society: the one is utility and the other is culture. Education must enable a man to become more efficient, to achieve with increasing facility the legitimate goals of his life.

Utility and culture are two recurrent themes that are particularly relevant to vocational education and, therefore, of significance to construction education. Giving people the means by which they can be effective contributors to society is a modern, utilitarian view of education and one often advanced by the construction industry. Contrast this view with the traditional view of universities and their purpose as expressed by Cardinal Newman in 1856 (in Kelly, 2001):

> that the true and adequate end of intellectual training and of a university is not learning or acquirement, but rather is thought or reason exercised upon knowledge, or what may be called philosophy.

This more esoteric view of universities and their role in education has its roots in the eighteenth century study of classics and philosophy. At this time, access to a university education was restricted to a privileged few – the sons of the powerful and wealthy. Is this important? Not directly, but indirectly it may be. The development of universities worldwide has followed a similar pattern, with the age of an institution being related to its standing and tradition similarly dominating over innovation.

On the issue of utility, construction education has come to universities late in this development. There were few programmes in construction management before 1950 and those that did exist were generally options within some other major discipline such as architecture or engineering. The first independent schools of construction can be seen developing through the 1960s and into the 1970s. Most, if not all, are in the newer and, often considered, lesser universities. There has been a desire to justify the existence of construction programmes in a university environment; in fact, a desire to demonstrate that they are academically worthy. This process of development has led to two interesting outcomes:

- The curriculum of construction programmes has been packed with what are considered to be more academic subjects such as mathematics, law, economics and science; and
- The learning process has mirrored that of the traditional engineering subjects with the emphasis on lectures and teaching.

The drive towards a more academic position has been unceasing, and has led to universities now arguing that, in construction as in engineering, they provide an education and it is for the employers of their graduates to provide the training. Paradoxically, no one would argue that construction education is not vocational, but nearly all see themselves as providing only one-half of what such an education requires. Utility has been left far behind.

There are some critical comparisons to be made here with other sectors that have developed differently and arrived in very different places. Medicine, for example, is found in many older universities and its principal learning objective is to produce effective practitioners (doctors, nurses, etc.). Schools of medicine see their role as producing, amongst others, surgeons who not only understand the procedures, but who can also carry them out. Architecture has its roots in the study of building design as an art form and many schools are still to be found within faculties of art. Here the emphasis is on the practice of art as much as the theory. Painting, sculpting and playing music are fundamentally 'doing' things and follow similar patterns. It would be a strange school of architecture that produced a designer who couldn't draw. However, the comparison breaks down when the length of programmes is taken into account. Most of these courses with highly practical outcomes are longer than the standard undergraduate programme.

Interestingly, this issue is mirrored in the view industry has of its academics. The very best academics associated with medical surgery are to be found in operating theatres, demonstrating pioneering surgical techniques on real patients in front of students. Witness the work of Professor Magdi Yacoub on heart transplants in children. In architecture, when confronted with an academic, most practitioners would ask to see their design work and make a judgement based on the practical evidence. However, in construction management, academics are often seen as remote figures engaged in activities that are of interest to them, but of little value to practice. These are sweeping generali-sations and no doubt everyone will have examples that demonstrate that they do not apply universally. But there can be no doubt that the connection between academia and practice is much stronger in medicine and architecture than it is in construction and engineering. The reason for this difference could be a cultural one.

On the issue of culture, society often sees education as an instrument of change; certainly this is what King (1948) was hoping for. In construction, there has been a lot of innovation and development in recent times, both in terms of technology and policy. However, studies have shown that there is a systemic inertia within senior management to new ideas and that graduates are often frustrated by their inability to apply their knowledge (Fortune & Lees, 1993). If the senior managers are responsible for training new recruits, it is not surprising that the recruits become inculcated with traditional ways of doing things.

Construction education has been developing in a particular way, but the questions remain: Has this been the right way and where should it go from here?

The education supply chain

The concept of the education supply chain is new and needs some explanation and definition before an examination can take place. Supply chain management is part of the innovative management agenda and was developed in Japan in the automotive industry. Its use generally in the manufacturing sector in the USA and Europe can be traced back to the 1970s and 1980s, but you have to wait until the 1990s to see evidence of supply chain management being used in the construction sector. Even now, this is considered to be a new area for the built environment and the benefits of adopting supply chain thinking are still being developed.

Supply chain management is about getting the whole of the supply chain to align behind the needs of the client/customer. The key drivers are improving quality, func-tionality and value, while at the same time reducing costs. Where there is a clear vocational education structure that provides the main source of new entrants to the sector then this can be argued to be part of the supply chain, and as such it is the ed-ucation supply chain for that industry. This vocational education structure will have many parts including universities, colleges and in-house training schemes.

It is important to establish the value of the construction sector to the global economy both in terms of its output and the employment it provides. The amount that a country spends on construction is closely related to its income. In 1998, expenditure varied from US$5 per head in Ethiopia to almost US$5000 in Japan. This means that construction output, by value, is heavily concentrated in the rich, developed world. The high-income countries of Europe are responsible for 30% of global output, the USA for 21% and Japan for 20%. China, despite its huge size and rapid economic growth in recent years, lags a long way behind with only 6%. India has 1.7% (International Labour Organization, 2004) (Figure 14.1).

The distribution of construction employment is almost the exact reverse of the distri-bution of output. While three-quarters of output is in the developed countries, three-quarters of employment is in the developing world. Official data suggest there are

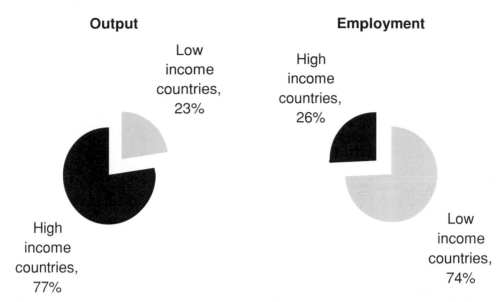

Figure 14.1   Output and employment between different economies (International Labour Organization, 2004).

around 111 million construction workers in the world, some 80 million of them in the low- and middle-income countries.

It is useful to recognise that the education supply chain exists within this global marketplace for construction education and, therefore, there are demand-side and supply-side issues to look at. On the demand side, the main issue is: What is the trend for new graduate recruitment? On the supply side, the issue is: Is the demand for new graduate recruitment being satisfied in terms of the quantity and quality of new recruits?

The demand for graduate recruitment

The developed countries with higher labour costs use more technology and higher level skills in construction to generate higher levels of output with fewer people. The less well-developed economies with lower labour costs still rely on low-level technology and highly labour intensive practices. In both circumstances, there is a demand for professionally qualified people to perform and manage the processes in construction. There is a close correlation between output in the sector and the demand for new recruits, and, therefore, new graduates from construction related programmes. Growth in construction activity has been consistent during the 1990s and early 2000s. A study by Davis Langdon and Seah International (2004) shows strong growth projections for the construction sector (Figure 14.2). In the UK, output from the construction sector rose by 23% in real terms during the period 1995–2003 (Figure 14.3). In the USA, a similar pattern exists; while Europe and the Pacific Rim areas have not shown quite

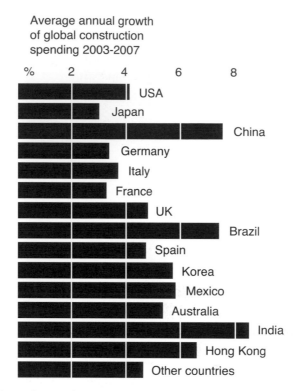

Figure 14.2   Estimated growth in construction.

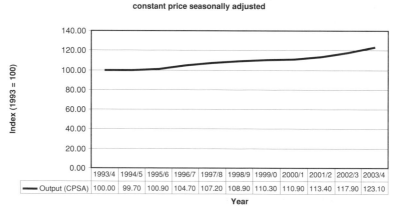

Figure 14.3   UK output (DTI (Department of Trade and Industry), 2006).

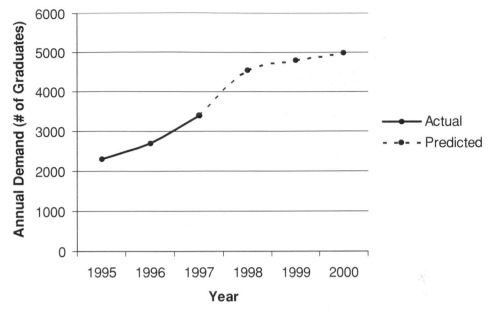

Figure 14.4   US demand for graduates (Bilbo et al., 2000).

the same consistency of growth, output is now significantly higher than that in the early 1990s.

This means that the demand for new recruits is growing and is projected to continue to grow. In the USA, Bilbo et al. (2000) identified that the demand for new graduate recruits more than doubled during the period 1995–2000 (Figure 14.4).

The developing countries will experience substantial growth in construction activity as the demands of an ever-increasing industrialisation process are brought to bear on the infrastructure. Presently, construction in these countries is highly labour intensive, but competition is likely to drive improvements in productivity and this will be facilitated by improved processes and higher level skills. This growth will, therefore, be both quantitative and qualitative in nature.

Examining the level of demand by looking at national output statistics and trends is useful, but it must be remembered that the real demand is related to activity and this occurs at a local, project level. Where there is major infrastructure development (such as Heathrow Terminal 5 or the new airport for Hong Kong), there can be a substantial rise in the demand for new recruits in that area thus creating a hotspot. Similarly, when the major projects have been completed, there can be a substantial drop in demand for new recruits thus creating a weak spot for employment.

To summarise the demand-side position:

- There is consistent growth in the global construction market.
- The higher level skills, and therefore the demand for new graduates, are currently concentrated in the developed countries.

UK - full-time graduates

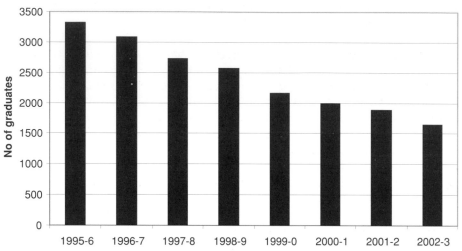

Figure 14.5 Number of full-time graduates from undergraduate programmes (HESA, 2006).

- There is potential for huge growth in the construction education market in developing countries.
- The demand for new recruits follows the construction activity and can cause local hotspots.

The supply of new recruits to industry

Bilbo et al. (2000) concluded that 'construction programs are not meeting the current demands of the construction industry [in the US] and it is unlikely that this fact will change in the projected future'. This conclusion was drawn because, while the supply of new graduates was expected to remain static, the demand generated by growth in activity would mean that more graduates would be required.

The position in the UK is in many respects worse than that in the USA, as the supply of new UK graduates from full-time programmes almost halved during the period 1994–2003 (Figure 14.5). There is still a substantial shortfall in the supply of new recruits to industry. Most full-time graduates in the UK and USA can make a choice from anything up to five employment offers. Moreover, some of those job offers come from other industries, such as management consultancy and finance, where the knowledge and skills of those with construction qualifications have been recognised and rewarded at a higher level.

The fall in applicants for full-time construction programmes would appear to be caused by a change in the way students choose their career direction and, therefore, their preferred university course. Traditionally, there was strong careers advice within schools and colleges and this was reinforced by parental guidance. Most recent research suggests that the way students currently choose their career is on the basis of 'perceived lifestyle'. This means that they choose on the basis of the kind of lifestyle, in terms of interest and rewards, they think they will have if they choose a particular career.

The old careers advice service is effectively moribund, so students have turned to alternative sources of information and, not surprisingly, the general media, particularly TV and radio, are the most significant in this respect. Teenagers enjoy a daily diet of programmes that focus on news, health, courtroom dramas and performing arts; and they play games using the latest developments in information technology. No one should be surprised that, as a consequence, the most popular programmes at universities are in media, journalism, law and information technology. There have been few positive images of construction, although there have been some exceptions for design and architecture; and it is no coincidence that interest in architecture courses has been reasonably strong while construction has been in decline. The best that construction can hope for is that the prospective students have no perception of construction, as if they do it is often negative – dirty, insecure, unsafe, low paid.

In the UK, this trend has been identified and its threat recognised. In the last few years a major advertising campaign has been run by the Construction Industry Training Board (CITB–ConstructionSkills) to promote understanding of the various aspects of construction and the diversity of careers within the industry. The campaign has been targeted at young people and has involved TV, radio and press advertising. This has been highly successful, with a substantial increase in the number of applicants to full-time construction programmes (33% increase in first year registrations from 2002 to 2004 – source UCAS).

The shortage of new graduates impacts on industry recruitment costs. There is also anecdotal evidence that these costs are increasing in a disproportionate way. It is now common that the money paid to recruitment consultants in order to secure an appointment is equivalent to 40% of the annual salary of the new recruit. Worse still, companies struggle to hang on to the staff they secure, with the very agencies employed to recruit staff being the source of the problem as they will maintain a relationship with the new recruit and inform them of alternative job offers when they become available.

A number of other changes can be observed that go some way to compensate for the shortage of new graduate recruits into the industry. Recruitment to part-time programmes has been growing in the UK and many other countries where there is a part-time tradition. In 2002, the number of students in the UK part-time undergraduate programmes exceeded the number in full-time programmes for the first time since construction management programmes became part of the university mainstream. There has been an increase in the number of conversion programmes where students, whose first degree is not cognate with construction, study a postgraduate programme in construction, often part-time, to enable them to take up employment in the industry. There is also a growing trend for those who missed out on formal education in their early careers to go back to college and improve their formal qualifications. And finally, there are some late career-changers, where people in their thirties and forties are making a complete career change from some other field to construction – largely because of the good employment opportunities.

This section has been concerned with the quantity dimension of education supply. To summarise the quantity issues on the supply side:

- The number of students studying full-time construction programmes has stayed constant or fallen in recent times.
- The construction industry struggles to recruit the best graduates and loses some to other industries.
- There is a substantial shortage of new graduate recruits to the industry.
- In some countries, there have been changes in the pattern of education in order to compensate for the shortage of new graduate recruits.

The quality of new recruits to industry

How many times have educators heard the words 'these graduates are not as good as they used to be' or 'I took on a new graduate and they couldn't do x or y'? Professional academics should judge their contribution to the industry by the value added by their new graduates. Or at least, this is how it should be in vocational education. The development of construction education has led many observers to identify a disconnection between industry and education (Williams, 2004). The evidence clearly shows that the education supply is failing to deliver the right number of new recruits, but what about quality issues? Is the education supply providing the right type of new recruits with the right mix of knowledge and skills?

The answer would appear to be no. The Construction Industry Council in the UK undertook a major survey of professional skills in 2004. This survey was aimed at those working in the main professions within construction and included designers, engineers, construction managers, project managers and surveyors. In parts of the survey, there was a focus on new recruits and the main findings for higher education are summarised in Table 14.1.

An ongoing research project in the University of Salford, has already identified the following preliminary findings (Lees, 2005). The study includes a survey of 164 construction graduates and their first employers (46 organisations) approximately one year after they took up employment. There were many positives to take from the study, with most graduates indicating that they found it 'relatively easy' to adjust to work, that finding a job was 'easy' and that their education was 'useful'. Employers were less convinced about the quality of education – citing disengagement and a

Table 14.1   Findings of the CIC Professional Skills Survey.

- 43% of respondents thought that the quality of new graduates had fallen in the last 5 years
- Skills gaps were identified in new graduates, specifically in:
  — technical and practical skills
  — literacy skills
  — communication skills

Table 14.2   Preliminary findings of study into graduate experiences on entering employment.

Quality of construction education

- 80% of employers and a staggering 84% of graduates think there is room for improvement in the programmes, with 79% of graduates indicating that they could have learnt more in their undergraduate studies.
- 65% of employers believe that students are not taught how to apply their knowledge.
- 43% of employers do not think that graduates have the appropriate skills for their chosen employment, whereas 35% of employers think they do have the right skills. This contrasts sharply with the view of graduates, where only 25% do not think they have the right skills and 54% think they do.

Education/industry engagement

- 46% of employers think there is little dialogue between them and education providers, with only 24% disagreeing with this view.
- 96% of employers think that industry and universities need to communicate more effectively.

lack of communication skills as serious issues. The preliminary findings are given in Table 14.2.

To summarise the quality issues on the supply side:

- The quality of graduates is perceived to be falling.
- New graduates lack practical, technical, literacy and communication skills.
- Employers have to invest an increasing amount of time and effort in training graduates so that they can become productive employees.

The skills debate – what should we be doing in the programmes?

The two studies in the earlier section take a UK perspective, but there is no evidence to believe that this problem is unique to the UK. So what is going wrong? There would appear to have been a divergence in the way universities, the professional bodies and employers have developed since the 1980s. This divergence has led to a disconnection between universities and industry, which has reinforced the idea that education takes place in universities and that training takes place in work. To answer the question of whether this divergence is in the interests of the industry or not, the whole issue of education and training must be revisited.

What does vocational education mean? This is a question that has taxed philosophers, researchers and practitioners alike through time, and it may be that there is still no consensus on this important issue. However, the purpose of education is a critical issue when considering the nature of universities and their role in vocational education.

The Greek philosopher, Socrates, argued that education was about drawing out what was already within the student. Indeed, the etymology of the word 'education' is from the Latin *e-ducere* – to draw out (*Webster's New World Dictionary*, 1956). The Sophists, by contrast, were concerned with providing their students with the knowledge to take up important positions of State. It would not be appropriate to attempt to resolve, or even debate fully, the philosophical arguments. But it is important to debate the issues of education and training and to do this some definitions, at least sufficient for the purpose of this debate, must be set down.

Education is about helping learners to develop their knowledge and understanding. Knowledge is the body of information about which something is known or believed to be true. The classical view of knowledge developed by Plato and Aristotle is that it can take three forms – comprehension of arguments; practical command of techniques; and wisdom to solve problems (Jonsen & Toulman, 1988). A more modern view is that it can take many forms: explicit; tacit; implicit; inherent; organisational; individual; and many more (Polanyi, 1964, 1966). Understanding is to do with meaning and insight and helps explain the way knowledge is used. For example, Newton's third law of motion states: '. . . to every action there will be an equal and opposite reaction'. As a piece of knowledge it has some interest, but no great value. The ability to give it as an answer in a quiz may provide some benefit to knowing it. However, it becomes really valuable when the understanding of what it means is added. As a person walks, they make forward motion by pushing their feet against the footpath; they rely on friction and the fact that the Earth is very large and heavy. Most times this reaction is sufficient and they make their way without incident. But if they try the same approach while standing on a skateboard, there is a different and less satisfactory outcome.

There is some trade-off between knowledge and understanding in that it is possible to consider all the implications as pieces of knowledge, a bit like a computer playing chess that considers all the possible moves before choosing the one that leads to the greatest benefit. Theoretically, if you have sufficient knowledge of all possible situations you do not need as much understanding. However, the value of understanding is in its ability to predict the outcome of situations that have not yet been experienced. Most people 'know', or believe, that being hit by a car when crossing a road would cause injury, although few have ever experienced it. They also understand the way in which cars make their way along roads and use this knowledge and understanding to work out when it is safe to cross.

The main focus of education is, therefore, the acquisition of knowledge and the development of understanding. Teaching and learning that is focused on education is full of information and explanations that can be presented to students through lectures, tutorials and seminars, or facilitated through other forms of teaching and learning like problem-based learning and action-learning.

Training is to do with skills and capability. This means the ability to apply knowledge in a practical or real situation – the ability to do something rather than just think about it. The focus here is on application rather than acquisition, although the latter is not ruled out. Training programmes can be characterised by the amount of practice they include. By practice, we mean the opportunity to perform a particular task repeatedly until it can be performed well on every occasion.

Vocational education incorporates both education and training. Where a particular profession is the objective of students, their development must include all aspects of knowledge, understanding, skills and capabilities relevant to the profession.

There can, however, be quite different approaches to vocational education. Here is an apocryphal story. Imagine that the purpose of a programme was to produce a qualified radiographer in a hospital, i.e. someone who takes X-rays of patients. In Russia, it has been said that on the first day of the course, students would be introduced to the X-ray machine and provided with an explanation of which buttons to press to take an X-ray. By the end of the first week, they would be able to take most basic X-rays successfully. At the end of a month, they would have completed their training and be qualified to work in a hospital. However, each time a new machine was introduced the radiographer would need to be re-trained.

By contrast, in the UK the first thing that radiography students do is to start a study of nuclear physics and anatomy. In their second year, they are introduced to various X-ray machines and at the end of year three, they graduate as qualified radiographers. Their more rounded education and training means that they can absorb new developments in radiography without significant investment in the need for re-training.

These two very different approaches to vocational education demonstrate the inherent tension between education and training. The Russian programme places more emphasis on training and achieves its goal in a shorter time. The UK programme probably produces more flexible and, in the long term, more valuable employees, but requires much greater investment up front.

The aim of most vocational programmes is to produce reflective practitioners (Schon, 1991). A key issue in the design of any vocational programme is the balance between education and training. The traditional view is that these two aspects are mutually exclusive, and that if you pack a programme full of training then there must be a corresponding reduction in the educational component. This view was challenged with the introduction of problem-based learning and similar application-based approaches to teaching and learning. While there have been some concerns expressed over the knowledge gaps that can surface in people who have learned through problem-based learning, overall the levels of competence achieved are higher.

The conclusion to be drawn from the success of problem-based programmes is that education and training are not mutually exclusive. Indeed, the development of knowledge and understanding with skills and capability, while being quite different, can arise from the same learning opportunity.

In the introductory section to this paper, the medical professions and architecture were given as examples of where the educational programmes in these disciplines are aimed at producing graduates with an appropriate balance of education and training; a balance that makes them competent at the point of graduation. This is not to say that they do not continue to gain experience or to develop and practise their skills – clearly they do. But what is equally clear is the design of these programmes is very focused on producing the finished article.

An examination of these types of programmes reveals another interesting outcome. In all these programmes there is, alongside the academic qualification that is the main purpose of the study, a statement of the practical competence of the graduate. In radiography, the graduate will attain a diploma indicating the classification of degree together with a certificate of registration that testifies to their competence in their intended career. In architecture, there will be the degree certificate, but there will also be a portfolio of designs that the graduate can show prospective employers to indicate their competence and provide the basis for professional qualification.

In most countries, and in nearly all construction education, there is no equivalent and separate indication of competence. This is due to construction education following the engineering paradigm. Education and training are separated, with education being left to universities and colleges and training being the domain of industry and the professional institutions. Many observers think that this is an unsustainable position that is detrimental to effective business. At a meeting of the Construction Industry Council Education Round Table, held in October 2004, in London, there was a call for a 'radical rethink' of construction education. The professional bodies were cited as having a 'silo' mentality that is restricting change in educational programmes and strangling real innovation.

The defence to this contention is often the different cycles to which industry and academia respond. Industry is more focused on 'now', while academia looks several years ahead. There is no doubt that employers know what they need now, but will it be the same thing in four years time? These different cycles will militate against a level of detail in the skill requirements, but there are key skills associated with certain professions that have varied little through time. Many people in the world of work see academia's insistence on a long-term view as an excuse for not engaging with the skills debate.

Figure 14.6 shows the relationship between education and training and sets out the current position. There is a need to improve the level of skills and capability that graduates in construction education have – in effect to move from Box 3 to Box 4. This can only be done by re-engaging with industry and working together to infuse programmes with the necessary context in which to develop the skills and capabilities. At the same time, it will be appropriate to raise the question about certificates of competence and whether these should be introduced to indicate the level attained by students at the end of their degree programme. These certificates would not need to be prescriptive, as they are in the field of medicine, but could be elective. They would contain only those skills and capabilities that the student had attained on the particular course. As such, they would aid the communication between potential employer and graduate in the same way as the portfolio does for an architecture graduate. Figure 14.6 also raises the issue of whether the training of trades-people should move towards Box 4? It could be that a movement in this direction would be helpful since it would support the increasing use of new technology and innovations. However, it could also be argued that the introduction of modern methods of construction produces a de-skilling of the workforce with the removal of traditional skills and crafts. In this case, a move towards a higher level of knowledge in the trades-people may not be helpful. The answer to this dilemma is beyond the scope of this chapter.

Figure 14.7 shows a way of reconciling the tension between the traditional academic approach of universities with the needs of vocational education. The diagram is a corrupted version of a Venn diagram using triangles instead of circles, while the vertical

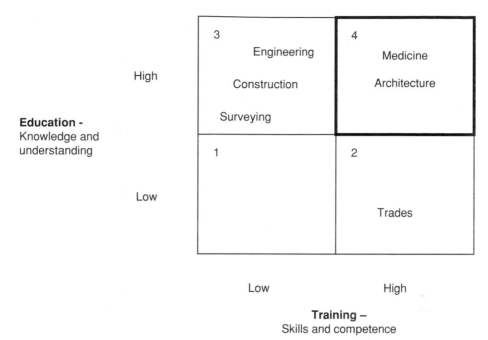

Figure 14.6   The learning quadrant.

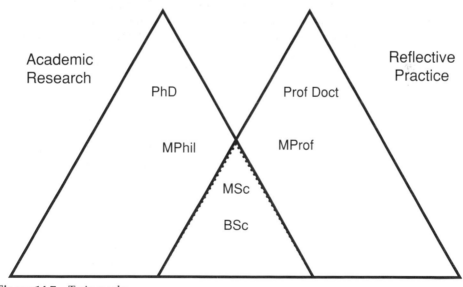

Figure 14.7   Twin peaks.

axis is used to indicate the degree of the challenge – the higher up someone is, the greater the achievement. It depicts two mountains that an individual may climb. Both mountains present legitimate objectives for an individual learner, but each mountain presents its own unique challenge. It may be that the techniques required to climb the academic research mountain are different from those for the reflective practice one, but the equal height of the peaks indicates that degree of difficulty and the reward for reaching the summit should be similar. The lateral disposition of the two mountains also says something about the current state of education and training in a subject. On the basis of the arguments already presented, the twin peaks for architecture and medicine would overlap substantially. The two summits would be relatively close to each other and the ridge that connects the two mountains at their highest interconnection would be only a short distance from the summits. For construction education a similar analysis would show a limited overlap and large lateral displacement. This would illustrate the disconnection between academia and industry/practice and go some way to explaining what must be done to put it right. The question follows: Can the mountains be moved closer together? If they can, it will be through greater engagement.

The way forward and the need for greater engagement

So what is the role of universities, is it to provide academic education or vocational training? Are they in the business of education, skills development or the new business of knowledge transfer (Ternouth, 2004)? Globalisation and the resulting economic and political pressures are demanding greater relevance from universities for the interests of national economic development and for the good of national economies. This has led to advocacy for a more integrated approach between government, industry and higher education institutes (HEIs) to attain maximum value from public and private investments. This resonates with the arguments in Chapter 4. Government, business and HEIs interact in a legislative and policy environment but also in this global economic context. Each has their own interests and values, and this determines the nature and intent of their interaction.

In a recent Council for Industry and Higher Education publication, Ternouth (2004) postulates that 'Economic success is about the ability to absorb and apply new technology and innovation', and this is about human resource capability; essentially the labour force and graduates of yesterday, today and, importantly, tomorrow. Can they exploit the opportunities that arise? So from the perspective of the learner, lifelong learning and developing the skills to achieve this are important as they prepare to enter the world of work.

A consensus appears to be emerging in the built environment education literature for a need to link the reality of the industry more closely with the learning and skills development of the student, whilst they learn. As Mole (1997) highlighted:

> Many of the criticisms recently about graduates are that they lack the basics and are unable to make the adjustment from university to practice. The debate, which revolves around what universities should be teaching and what is best left to practice, has become very apparent.

Williams (2004) illustrates the range of possibilities that exist in trying to achieve this in the Accelerating Change in Built Environment Education (ACBEE) study of case examples from the UK.

Mode 1 and Mode 2: Balancing the traditional and context-driven approaches

This need for greater engagement can be related to the development of the new educational paradigm and the shift from old to new methods of knowledge production. This categorisation of all forms of scientific knowledge into Mode 1 and Mode 2 was initially debated in the early 1990s, and resulted in two challenging texts by Gibbons et al. (1994) and Ziman (1994), followed up by *Re-thinking Science* (Nowotny et al., 2001) which challenges the established relationship between science and society.

This new educational paradigm is emerging thanks in part to both new technology and an increased focus upon the individual student and their particular learning and development needs (Hills & Tedford, 2003). It is interesting to note that these needs are addressed in an increasingly integrated manner, working in small groups. This approach can be summarised as the demise of the lecture and its replacement by the tutorial – the time-honoured procedure of Oxbridge and other institutions that can afford small classes. The old approach of formal lectures delivered in a lecture theatre will increasingly be replaced by small group tutorials, case studies and experiential learning opportunities supported by electronic resources and probably an electronic classroom. These developments increase the opportunity for part-time and distance education in the process.

The shift that is identified here can also be understood as the shift from Mode 1 to Mode 2 style learning; from single discipline, classroom-based and passive to inter-disciplinary, integrated, context-focused and experiential. Representing the traditional academic approach, Mode 1 is characterised by the following attributes (from Hills & Tedford, 2003):

- Fragmentation of knowledge
- Internal referencing, peer review, cronyism and social corruption
- Absence of context, flight from reality
- Objectivity taken to extreme, dehumanisation of science
- Authoritarian attitudes to knowledge and success
- Competition between knowledge bases leads to internal uniformity and external conformity
- Academic values prevail, theory prevails over practice

The alternative, which represents Mode 2, is based upon the contextualisation of knowledge, and is characterised by the following:

- Holistic and not reductionist
- Context driven, not subject driven
- Mission-oriented research, not blue skies
- Teamwork, not individual scholar

- Multi-authored publications and heterogeneous knowledge bases
- Divergent, not convergent thinking
- Reflexive philosophy rather than objective statements
- Decisive criterion: does it work?
- This is the world outside academia

Adopting a more integrated approach will facilitate this shift from Mode 1 to Mode 2 teaching and learning. In essence, Mode 2 could be viewed as Mode 1 operating in a contextual setting, dealing with the issues and the problems faced by the industry. This point has been noted by Hills & Tedford (2003):

> The reform of undergraduate education. . . has been concerned to shift the emphasis from didactic lectures, subject-centred teaching, the rote learning of facts, explicit knowledge, and memory based examinations to the softer world of student-centred learning, tacit, implicit knowledge and continuous assessment.

Graduate education in the built environment is embracing this shift from Mode 1 to Mode 2 context-based education, and examples of this can be identified in all the case studies undertaken in ACBEE phase 1 (Williams, 2004). The extent and nature of the engagement identified in the ACBEE study supports the vocational nature of construction courses.

There would also appear to be some scope in aligning the shift from Mode 1 to Mode 2 to the stage of the student's education. So, for example, a first year undergraduate is more likely to receive more Mode 1 style teaching than a final year undergraduate with work experience who could positively participate in Mode 2. Likewise, the professional in practice undertaking continuing professional development (CPD) activities would most likely engage in Mode 2 learning. An interesting point to note is that it can be argued that a doctoral student is likely to pursue a Mode 1 oriented set of activities, researching a particular area, often out of context, in the academic laboratory. However, other postgraduate students may embark on programmes based on Mode 2 learning. This approach supports the idea that undergraduate education should be of a general nature and postgraduate education about developing the specialist in areas such as law, design management, facilities management and project management (Love et al., 2001).

With the growth of action research and action learning style activities, more contextual studies and opportunities for engagement between individual students or groups of students with industry will be provided and will increasingly lead towards postgraduate qualifications. There is no doubt that technology will help with the adoption of a Mode 2 approach to education. Increasingly, it is difficult to take students on visits to major projects, due to health and safety considerations. The use of site-based web cams will allow the site to come to the university. Organisations, like Gehry Technologies, are already sharing their project information with schools and colleges so that they can see leading edge design and production.

The diversity of learners means that we will see a growth in the number of remote-learners who use technology to create a virtual community in which they can share their ideas. This approach will not be exclusive to remote-learners as more campus-based students turn to technology to connect with other students on different courses in different institutions. This will also blur the international boundaries, with learners, and tutors from across the world linking together to share their experiences. The push towards integrated teams for construction projects means that inter-disciplinary learning will become a must in undergraduate programmes. Information and software application sharing through web-based interaction enabled through programs such as HorizonWimba will be the way this is facilitated in the future. The campus of the future will not be virtual, but it will use technology to extend its reach and engage more fully with learners and industry/practice.

Barriers to developing engagement

The Mode 2 approach will entail closer engagement between industry and HEIs to provide the student with a more contextual learning experience. Currently, however, numerous barriers need to be overcome before such integrated practice can be considered the norm rather than the exception. The barriers to developing successful engagement identified in the literature include:

- High transaction costs in establishing such collaborations (Lambert, 2003; Ternouth, 2004): in particular, the difficulties which exist in the negotiations surrounding intellectual property rights and complications which can arise as parties seek to protect their own self interests
- The lack of a common language, mistrust and the existence of a knowledge gap (Matlay, 2000; Luna & Velasco, 2003)
- Differing objectives and time scales for delivery (Valentin, 2000)
- Cultural differences between industry and academia, particularly small businesses (Tilley & Johnson, 1999; Matlay, 2000; Valentin, 2000)

In addition to the above, it would appear that the business case has not been made or fully established to support the establishment of such types of engagement. Despite this, there are examples of good practice, as illustrated within the ACBEE initiative (Williams, 2004). The business of engagement ultimately operates in a market, and collaboration will only continue and thrive if objectives are achieved and returns are seen to be positive.

Conclusions and future trends

This examination of construction education has found the following trends:

- Construction activity is growing and is likely to continue to grow in the medium-term.
- Developing countries especially will need more and more graduates to satisfy the needs of the construction industry.
- The number of new graduates being produced is insufficient to meet the demand.

- The quality of graduates falls short of expectations.
- There is little integration between students of different disciplines

This chapter concludes that several points need to be addressed to achieve the way forward:

- The image of construction as an industry in which to work needs to be substantially improved. This will generate much greater interest in studying full-time programmes in construction.
- A new paradigm of engagement between academia, industry, professional bodies and government is needed. This will allow an integrated approach to education that is better informed by strategy and less driven by the requirements of individual isolated stakeholders.
- Construction education providers should re-visit their curricula and embrace the lessons learned in other disciplines on the balance between education and training. This will lead to Mode 2 learning and an improvement in the quality of graduates, so they are more able to make a valuable contribution to the industry
- There needs to be greater use of technology to facilitate inter-disciplinary learning opportunities. This will aid the use of Mode 2 learning and produce more holistic practitioners.

The purpose of this chapter was to identify the long-term educational implications of revaluing construction. Inevitably, there is a focus on the things that are going wrong or are not working. This, of course, underplays the extremely valuable contribution to the industry made by the education providers. This good work is recognised, but the emphasis in this paper is: 'How can education be improved?'

A revaluing of construction must focus to some extent on the industry's presence in the perception of society. That is not to present an image that is untrue or misleading, but to present a view that is clear and appropriate and which measures properly the value that construction and the built environment make to the life of people. In recent times, the industry has allowed its true value to become hidden, and this has had far-reaching consequences for education and the supply of new recruits. All stakeholders should join together and accept their responsibility to put this right.

15 Revaluing construction: a building users' perspective

Jacqueline Vischer

Summary

This chapter builds on Chapter 8 by focusing on the pivotal issue of value to users. After all, buildings are built for people to use, but few methods and procedures exist for ensuring that the users' perspective is incorporated into building decision-making. Most design and construction decisions involve trading off building quality against construction cost. Cost is usually defined in a short-term, one-time framework with little consideration of the long-term impact of cost-cutting decisions. As a result, quality for users is often compromised. In spite of efforts to systematise pre-design briefing and post-occupancy evaluation, this cost-oriented perspective, highlighted in Chapter 11, typically extends through all stages of building delivery and occupancy. An alternative approach known as 'building performance evaluation' is based on feedback loops at each stage of building delivery and throughout occupancy. The feedback is generated by users and stakeholders for systematic incorporation into decision-making. Feedback on users' experience of buildings needs to be structured around a functional theory of building occupancy. Classifying building users' experience into the three categories of physical, functional and psychological comfort provides a useful framework for collecting and analysing user feedback. The users' perspective can then more easily be fitted into the feedback loops that are useful to design and construction decision-making.

Introduction

It is sometimes hard to remember that buildings of all shapes and sizes are fundamentally built for one reason: because people are going to occupy and use them. Relatively few buildings are constructed for no human occupancy (telephone switch centres, barns and warehouses, Lenin's Mausoleum, are some examples). In this chapter, we examine the apparent paradox that people almost never have a say in the buildings they will be using, that their interests are routinely discounted in favour of other priorities of a technical and financial nature, both during the construction of new buildings and during the lifetime of the structure, and that major research efforts and knowledge-gathering over the years have failed to influence significantly the way building design and construction decisions are made.

Some of the ways in which the construction industry has historically failed to meet its obligations towards the ultimate consumers of its products are a result of financing requirements. The priorities that make most sense to banks and other financing institutions are accepted at an early stage. Unfortunately, these criteria are not necessarily those that ensure quality in the final product, as they tend to favour solutions providing a short-term return on investment rather than solutions that provide a high-quality environmental experience for building users. Capital funds are loaned under conditions that favour speed over thoroughness, are limited to the short-term time-frame of building delivery without reference to the long-term considerations of building operation, and have the effect of discouraging innovation and risk at all levels. The conflict between long-term – but not proven – environmental quality objectives, and short-term tangible returns on investment is played out at every stage of building decision-making. Little data are yet available on the financial and social costs to society of building and occupying buildings that do not work for users and that ignore issues of habitability and sustainability.

This financially driven perspective means that conventional building practices are often preferred on projects, creating an industry-wide culture that inhibits innovation because it is risk-averse. Innovative materials and technology are hard to justify financially before they are 'proven', thereby discouraging builders from experimenting with innovation even though conventional practices may lead to outcomes that are less effective, comfortable, or adaptable once the building is completed. Quality in buildings tends to be defined more by marketing criteria than in terms of the level of comfort and functionality of the spaces people will occupy. In office buildings, for example, more tends to be spent on public spaces that have no real function – such as large high atria with expensive marble and bronze finishes – than on narrow office floors with high ceilings that ensure daylight for everyone, as well as screen-based work-friendly indirect artificial lighting; or on high-grade mechanical system solutions that allow users both to open windows and to receive good quality ventilation for a variety of office and workstation layouts.

In a recent seminar on productive work environments, an engineering firm described two options for renovating and upgrading an ageing office building near London. One alternative allowed the windows to remain operable while sheathing the building envelope in a glass 'skin' that ensured daylight penetration while at the same time controlling heat gain and providing cooling. The other option required sealing the building (specifically, the windows) and installing a conventional air conditioning system. While the first solution increased user *comfort* by offering control over window operability as well as user *satisfaction* by vastly improving the building's appearance, the second solution was the one selected owing to the relative ease of financing and marketing an 'air conditioned office building'. The front-end costs of both renovation options were comparable, but the vastly different effect on users and on the long-term liveability of the building did not apparently enter into the calculation.

Innovative (read, 'improved') building design often stops short at the level of architectural design. The new 'gherkin' on London's skyline, the new 'tallest in the world' structures being built in Shanghai and Tokyo, the proposals for the World Trade Center site in New York City – to name a few contemporary examples of innovative buildings – are more oriented to form and appearance than they are to improving the environmental quality of interior spaces occupied by users. Efforts to innovate at the level of building systems and technology performance and energy management, for example, where the impact on users is likely to be more pronounced, are discouraged both by the low visibility of such improvements (they are hard to 'market' because they cannot be seen), and by the reluctance of client and tenant decision-makers to pay higher rents for a better quality work environment for staff. In a recent example in Montreal, Canada, one of the downtown's newest buildings is a 'high-tech' intelligent office tower where the developers, seeking high-profile, high-tech firms as tenants, installed underfloor 'task' ventilation. The extra cost obliged them to offer leases at CAN\$2–4 per square foot more than other Class A office buildings. The relatively high vacancy rate meant that they had to market the building's innovations and increased 'habitability' in order to compete, but these features were not attractive enough to convince most tenants to move into their more expensive building, causing the developers to regret their 'investment' in this innovation.

As a result, therefore, of the way the building industry operates and is seen to operate, commercial buildings are often inflexible, costly to run, uncomfortable, not ecologically responsible and over time require significant reinvestment to 'make them work'. This process, which has been characterised as dysfunctional by many writers, is in part due to the fragmentation of the building professions and the lack of informed decision-making at every stage. Traditionally, the architect or master-builder exercised control over the entire building process, and could, by implication, ensure a high-quality result. However, the architect is now one of a series of specialists involved in a complex and expensive project over which no single participant exercises control, including the relatively 'new' professions of project management, facilities management, interior design and space planning.

Quality and cost

Because the building industry as it currently operates in most countries is largely cost-driven, cost calculations are applied to every stage of construction, and 'time is money'. The opportunities for gathering and applying information about users to building decisions are therefore limited. A large and exhaustive array of books and publications on briefing or programming bear witness to the obvious value of systematically consulting users and recording their requirements. But ignorance still persists in the construction industry not only on what briefing/programming is,[1] but also what it could be, and how to carry it out (Barrett & Stanley, 1999).

Some projects, such as office buildings, are built to preset standards to meet zoning requirements and respond to market demand. Briefing is not considered necessary in these cases as long as the amount of 'rentable' or 'useable' space is maximised. On buildings with more complex functions, such as hospitals, briefing is confined to technical performance requirements and systems specifications. Information-gathering initiatives such as briefing and post-occupancy studies, as well as using sustainable

---

[1] The UK term 'briefing' will be used in this book. In the USA, for example, this should be read as 'programming'.

materials and processes, take extra time if they are 'added on' incrementally to a pre-existing, linear building delivery process. Most building project cost-estimating is based on a series of sequential steps based on estimates of time needed to complete them. Thus, using conventional evaluation criteria, taking the time to acquire more or better information, seek feedback from stake-holders, and otherwise inform decisions adds to the length of time needed until the building is built and occupied and starts generating revenues, and thus to the cost of the financing needed for the project,.

Several years ago, this author wrote, 'Environmental quality is that combination of environmental elements that interact with users of the environment to enable that environment to be the best possible one for the activities that go on in it' (Vischer, 1989) For building developers, quality in buildings can mean concessions to a higher level of comfort for occupants than clients may be willing to pay for. In a more user-oriented building industry, quality would not be an extra that clients have to buy, but rather a basic responsibility that building providers and managers have towards the users of buildings. Replacing *time* and *money* with *quality* at the centre of the building construction and operation process would do away with the short-term, cost-driven, piecemeal approach to delivering and managing buildings, and, therefore, with the added costs of managing buildings resulting from this process. It would also do away with the notion that taking time to acquire information and feedback to inform decisions is not cost-effective.

Briefing is a crucial tool in the process of defining building quality, yet tying design and construction decisions too strictly to current client/user requirements can also en-gender long-term costs (Blyth & Worthington, 2001). It is important to remember that flexibility is an important ingredient of building quality. As modern office buildings built in traditional ways lack the ability to adapt easily and cheaply to changing functional demands, this lack of flexibility adds to operating costs. But the decisions that would have made a building 'flexible' are made while a building is being planned, designed and financed and not after it is occupied. Decisions that increase building quality by providing flexibility are not usually taken at the appropriate time, either through ignorance, or because a cheaper option is chosen. Flexibility is that building property that enables the organisation's activities to be supported into the future, even as technology and business processes are changing. Flexibility thus has value to the occupying organisation, and a lack of it increases operating costs.

If the building industry does not always, and everywhere, build the cheapest building with the cheapest materials, it is because clients and tenants attach value to quality. This goes beyond the value for an individual user: the meaning of 'quality' for facilities managers is related to the needs and requirements of their clients and of the building users who are their customers. Through being attached to and involved with the organisation, facility managers usually come to know what their customers need, and they translate this into quality criteria (such as service level agreements) that they can monitor. A major part of the facility manager's task is to: drive down energy costs; 'churn' costs incurred from reconfiguring workspace in order to move people around, and maintenance and repair costs; and to balance operating plus capital costs with possible income generated by leasing out space. Facility managers benchmark their costs against those of other companies and buildings using preset performance indicators that provide an assessment of whether the building is efficiently run. Theirs is a cost-driven definition of quality, which often fails to address other definitions of quality, or even other costs (Van Wagenberg, 1997).

Unanticipated costs arise as a result of the separation between construction financing and operating budgets that is built into the traditional approach. Thus all repair and renovation work that must be done after a building is occupied – to correct performance problems, to increase occupants' comfort and to accommodate changing uses and needs – is not calculated as part of the cost of delivering the building, but considered as operating costs. Moreover, if the building systems do not function as expected, the cost of time lost by occupants who cannot use a space or a piece of equipment is rarely added to the cost of repair or replacement. This 'downtime' signifies the negative economic value of disruption to work processes caused by the facility infrastructure not functioning properly. Applying knowledge of how people use buildings, not only in terms of requirements as determined by briefing, but also by placing the user's experience at the centre of the building delivery/procurement and management process would create a more reasonable balance between quality and cost.

Post-occupancy evaluation

An important tool that can be applied to all stages of design and construction has been developed to generate more knowledge of how people use buildings. Post-occupancy evaluation (POE) is the process of evaluating buildings in a systematic and rigorous manner after they have been built and occupied for some time (Preiser et al., 1988). POE started with one-off case study evaluations in the late 1960s, and progressed to system-wide and cross-sectional evaluation efforts in the 1970s and 1980s. Early POEs focused on the residential environment and subsidised housing, especially as a result of rapid home construction after World War II. Many urban renewal projects in North America, and New Town construction in western European countries, built large quantities of housing with little knowledge of the needs, expectations, behaviour or lifestyles of the people who would live there. The kinds of social and architectural problems that arose led to an interest in systematic assessment of the physical environment in terms of how people were using space (Vischer, 1989). This approach was later seen as a mechanism for collecting useful information for the building industry on the impact of design and construction decisions over the long term. POEs have since targeted hospitals, prisons and other public buildings, as well as offices and commercial structures.

Evaluation and feedback during the planning, briefing, design, construction and occupancy phases of buildings is often limited to technical evaluations related to materials, engineering or construction practices. Commonly applied evaluation procedures include structural tests, reviews of load-bearing elements, soil testing and mechanical systems performance checks, as well as post-construction evaluation (physical inspection) prior to building occupancy. POE differs from these in several ways, and is considerably less common. POE addresses the needs, activities and goals of the people and organisation using a facility. In some cases, POE results are published and widely disseminated; in others, they are uniquely available to the architect, to the client or to the stakeholder who commissioned the study. The findings from POE studies, while primarily focused on the experiences of building users, are relevant to a broad range of building design and management decisions. Many building problems identified after occupancy have been found to be systemic: information the engineer did not have about building use; changes that were made after occupancy that the architect did not design for; or facilities staff's failure to understand how to operate building systems.

POE in the UK

To date, POE has focused primarily on users' experience of the performance of buildings; the most recent step in the evolution of POE is one that emphasises a holistic, process-oriented approach. This means that not only the buildings in use, but also the forces that shape them (political, economic, social, etc.) are taken into account (Preiser, 2001). POE in the UK first emerged during the 1950s and 1960s. In its 1963 *Plan of Work for Design Team Operation* (RIBA, 1963), the Royal Institute of British Architects included a final Stage M ('Feedback') in which architects examined the results of their design decisions. Although this was later removed, building feedback began to be collected systematically by the Building Performance Research Unit at the University of Strathclyde. In 1999, a new call for 'systemising feedback and instituting post-occupancy evaluation' was issued, and in 2003, the RIBA Practice Committee re-introduced POE into its published documents (RIBA, 2003).

In the UK, as in most other countries, POE continues to be the exception rather than the rule. In part, this is because conventional building project stakeholders are not paid to perform POEs, because disseminating and applying the results is difficult and time-consuming. Another reason is because effective POE has to a tread a fine line between generating useful new knowledge about building use and criticising building professionals whose building decisions are often made in circumstances that have changed by the time the building is built and occupied. However, whereas in most countries POE remains an academic or research exercise, or else a private and 'one-off' initiative whose results remain confidential, PROBE (Post-occupancy Review Of Buildings and their Engineering) was established in the UK as a nationwide effort to solve these problems. Measurement included quantitative tools for 'hard' issues (the TM 22 energy survey method published by CIBSE in 1999) and a more qualitative tool, the Building Use Studies' occupant questionnaire, both of which have reference benchmarks. The results addressed not only the physical features of building design and construction, but also the activities of briefing, procurement, operations and facilities management (Leaman & Bordass, 2004; CIBSE, 1999).

After undertaking a number of studies and pooling POE results, the PROBE team concluded that cultural change is necessary in the building professions in order to incorporate feedback routinely into design, construction, procurement and management practices. It is difficult to get clients to adopt and pay directly for building feedback. In addition to a lack of interest by top management, who are rewarded for rapidly delivered buildings at or below budget, many clients know little about the range of techniques available, how they should be used and what value they might add. Clients who carry out post-completion checks to ensure that buildings comply with their requirements often have property portfolios with procurement departments oriented to individual projects, much like most designers and builders. As a result, they seek project sign-off checklists rather than real engagement with the in-use performance of buildings. Effective briefing also affects the feasibility and value of POE. PROBE found that client requirements are not always stated clearly enough at the outset to be rigorously applied to post-occupancy verification. Typically, briefing is reduced to a minimum in conventional building projects to increase the speed of the delivery process. Even where requirements are clear initially, they may change during the course of the project; if changes and their consequences are not formally incorporated into revisions of the brief or drawings, the initial briefing document loses its value.

Some feel that the weakness of POE to date is the failure to develop a consensus around the methods of data-gathering and analysis, to manage effectively the data collected, to identify benchmarks and comparators and to disseminate results without compromising the intellectual property of the service providers. These problems can be addressed by a more process-oriented approach to POE. A broader and more sweeping approach, known as 'building performance evaluation', places POE at the centre of building design, delivery and operations (Preiser & Vischer, 2004).

Building performance evaluation (BPE)

The BPE framework was developed in order to broaden the basis for POE feedback; it enables POE to be relevant earlier in the design process and to be applied throughout building delivery and life-cycle. The goal of BPE is to improve the quality of decisions made at every phase of the building life-cycle. Rather than waiting for the building to be occupied before evaluating decisions, ongoing feedback cycles provide information designed to inform key decision-makers. This helps avoid common mistakes caused by insufficient information and inadequate communication among building professionals involved at different stages.

BPE is based on feedback and evaluation at every phase of building delivery, ranging from strategic planning to occupancy, and covering the useful life of a building from occupancy to adaptive re-use/disposal. BPE came into being as a result of knowledge accumulated from years of post-occupancy studies of buildings, the results of which contained important information for architects, builders and others involved in the process of creating buildings – information that is infrequently accessed and even more rarely applied in most building projects. How then to systematise not only the research needed to acquire appropriate feedback from users, but also to ensure that such feedback is directly applied to the building delivery process, such that a response to it is incorporated at every stage? BPE is a way of systematically ensuring and protecting the quality of buildings while they are being planned and built, and, later, occupied and managed (Preiser & Schramm, 1997).

The BPE process with feedback through ongoing evaluation can be conceptualised as a series of loops through which information is fed back as continuous evaluation, leading to better informed design assumptions and, ultimately, to better solutions. In using such a process, stakeholders are able to make better and more user-oriented decisions as they are able to access the information gathered in evaluative research and stored and updated in information systems. The theoretical foundation of BPE has been adapted from the interdisciplinary field of cybernetics, which is defined as 'the study of human control functions and of mechanical and electronic systems designed to replace them, involving the application of statistical mechanics to communication engineering' (Von Foerster, 1985). A systems model is appropriate in building construction because it holistically links diverse phenomena that influence relationships between people, processes and their surroundings, including the physical, social and cultural environments. Like any other living species, humans are organisms adjusting to a dynamic, ever-changing environment, and the interactive nature of the person–environment relationship is usefully represented by the systems concept. Specifically, the systems approach to research in construction studies the impact of human actions on the physical environment – both built and natural – and vice versa. BPE has built on this tradition: it is multi-disciplinary and it generates the kind of applied research studies that, as diverse applications of POE, lacked a coherent theoretical framework.

As previously stated, applying BPE means that each phase consists of one or more feedback loops, such that relevant information from stakeholders is sought and applied to important building decisions. This must be done effectively without increasing project costs. Real-world examples indicate that the costs of such projects are less than might have been expected. And why? Because rational decision-making based on the right information at the right time was built into each project. The BPE approach ensures that feedback from stakeholders – not just from the building's occupants, but also facilities management, client/owners and lending institutions, as well as professionals in the building industry – informs each stage of the process in a systematic way. It is important to note that none of these phases, feedback loops and information tools are new. What is new is putting them together and presenting them as a whole and holistic approach to restructuring the way buildings are built and managed. The BPE approach is a comprehensive and overarching methodology of ensuring that a range of issues related to building quality are not just taken into account separately and occasionally at the whim of the client or the project manager, but actually drive the process.

BPE phases and feedback loops

The six phases and feedback loops of BPE are shown in Figure 15.1. In Phase 1, strategic planning is implemented using different types of feedback in the form of an effectiveness review. At this stage, clients make the first, critical decisions about their move, renovation or new building. The data gathered at later phases in the project indicate that investing time in strategic planning saves time and money later on. Conversely, a wrong or bad decision at an early stage creates problems and costs throughout the rest of the process. A strategic planning process can be structured to include all stakeholders, including diverse user groups, as long as the process is explicit and roles and responsibilities are clearly outlined.

Phase 2 of BPE is briefing/programming. This requires a thorough and systematic analysis of users' needs, resolution of potential conflicts, establishment of priorities and enumeration of performance criteria. Inadequate briefing increases project costs by failing to resolve or even to identify conflicts that emerge at a later stage, requiring

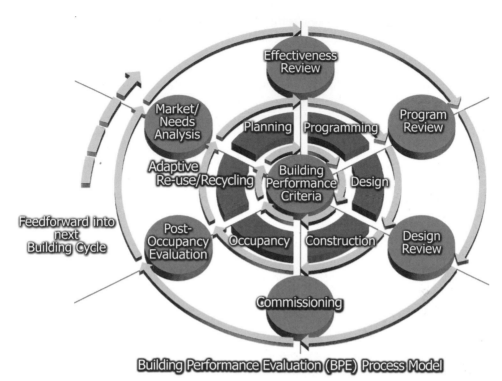

Figure 15.1   Building performance evaluation (BPE) process model. (Reproduced with permission from Preiser, W.F.E. and Vischer, J.C. (Eds.) *Assessing Building Performance*, Oxford, Elsevier, 2005.)

design revisions, change in orders during construction and changes and renovations after occupancy. A variety of briefing/programming methods and techniques have been published, mostly oriented to identifying stakeholders, cliciting their input and finding ways to set priorities, resolve conflicts and reach consensus on project objectives (Sanoff, 1977; Preiser, 1985; Marmot et al, 2004). If briefing/programming is effective, users and other stakeholders move smoothly through Phase 3: design and design review. Depending on how it is structured, design review is another opportunity to engage users in design decision-making in a valid and constructive way that enables information to feed forward into the next cycle of decision-making. A managed participatory process at this stage may appear to take longer, but ultimately results in a high-quality building that costs no more than a conventional project because there are fewer delays and less disruption as the process advances.

A commissioning methodology for Phase 4 – construction – is designed to ensure that the client/owner gets the building that was expected and anticipated. Again, if implemented correctly, feedback – from testing and inspections, from referring back to briefing criteria and from checklists and guidelines – is increasingly sought as a building project nears completion and moves towards occupancy. As with the previous phases, if planning, briefing and design review have been performed well, discrepancies between what was specified and what has been built should be minimal. Once a building is built and occupied, other tools are available to optimise building operations and ensure a quality experience for users. The principal tool for Phase 5 – occupancy – is post-occupancy evaluation (POE), and this type of feedback is sought by facilities managers during this, the longest phase of the BPE process. Systematic feedback on building performance from users places the user at the centre of the decision-making process, thus supporting and enhancing the role and responsibilities of facilities managers.

Phase 6 of BPE – adaptive re-use/building recycling – signifies the end of the building's useful life. As a building ages, questions proliferate about the relative value of investing in upgrades and renovations, and, ultimately, the feasibility of mothballing and even demolishing a structure. Again, systematic evaluation of alternatives using techniques like KPI (key performance indicators) and the 'balanced scorecard' approach, allows all viewpoints to be represented and ensures that relevant information from a variety of perspectives is taken into account. Direct and indirect input from all stakeholders helps to make cost-effective decisions and resist political pressure to respond to only one interest group's perspective.

To make sure the phases of BPE run smoothly and the feedback loops are implemented in a constructive and efficient fashion, key decision-makers, such as project managers, architects, clients or tenants' representatives and construction and facilities managers benefit from training in group process and facilitation. In addition, trained facilitators with their own skills and techniques can and should be employed to make sure the feedback loops are enacted appropriately, that consultations with non-professional stakeholders (users) as well as the involvement of all professional stakeholders are effective and constructive. Trained facilitation ensures that the process is cyclical and integrated, rather than linear with handoffs at every stage, and that it moves forward in a responsible fashion.

It is essential that BPE not be limited to functioning in only one culture or context. On the contrary, BPE offers a broad and adaptable framework for professionals affiliated with the building industry at all levels and in diverse cultures to find ways of implementing a user-oriented, cost-effective and high-quality approach to producing all types of buildings. In effect, BPE represents an optimistic and common-sense option for the future, the implementation of which can and will lead to better quality buildings and a better organised and functioning building industry.

Measuring performance through feedback from building users

In order for the BPE approach to work effectively, data-gathering and analysis activities are necessary at every stage. These can be carried out in a variety of ways, including, but not limited to, traditional social science research techniques. In view of the fact that performance criteria at each stage are constituted of both quantitative and qualitative performance evaluation, it is necessary to effect both quantitative and qualitative research. For instance, expected building performance in an area such as temperature levels inside a building can be compared with users' ratings of thermal comfort. For this comparison to be effective, both the expected and actual performance must use the same or comparable units of measurement. In some fields this can be complicated. For example, *expected* acoustic performance is usually given in the form of construction or materials specifications – distance on the centre of wall studs, sound absorption coefficient of a ceiling tile – but *actual* acoustic comfort of users is mostly received in the form of a satisfaction rating, in spite of the fact that users' satisfaction levels are often attributable to complex issues that go beyond building performance.

One of the challenges of the BPE approach is therefore to encourage more precise measures of user comfort than are conventionally used. Asking people whether they are satisfied is a rather broad and general outcome measure that tends to include far more than the performance criterion under consideration. Vischer (1989, 1996, 2001) has proposed a technique for approaching users with more direct questions about their comfort levels in relation to various building systems, in order to derive a more specific equivalent to objective environmental measures. The Building-In-Use (BIU) assessment system is a validated and reliable standardised survey that can be administered to occupants of any office building in order to collect simple reliable measures of their comfort in regards to key environmental conditions. From the responses, scores on each comfort dimension can be calculated and compared to a typical or average office building standard derived from a large database of user ratings that provide 'norms' against which building scores can be assessed. Thus, deviations from the norm for each building condition, in both a positive and a negative direction, provide a quantitative rating of what is essentially a qualitative measure.

The original seven building performance dimensions addressed by the BIU assessment are: air quality; thermal comfort; spatial comfort; privacy; lighting quality; office noise control; and building noise control. More recent results indicate that the comfort of the modern office worker, who must access a variety of equipment and perform a variety of tasks, additionally incorporates a perception of comfort in terms of security, building appearance, workstation comfort and overall visual comfort. User feedback in this relatively simple form can be applied to briefing, to design decisions, to operating and to budgetary decisions. It can also be used by design professionals who seek a measurement of comfort in an older or about-to-be replaced environment or in a new building soon after a move, as well as by building owners and business managers who seek to determine a baseline comfort level for employees in their buildings (Fischer & Vischer, 1998).

One advantage of the BIU assessment system is that it permits measured and perceived levels of performance to be compared. In parts of the building where certain areas of comfort are lower than the standard, instruments can be applied to measuring conventional comfort parameters and determining whether or not objective standards are being met. The user feedback provides a diagnostic data point that permits a wide variety of follow-up actions. However, it is not always the case that a direct correspondence can be determined between user feedback and the data provided by calibrated measuring instruments (Vischer, 1993). This is not surprising, in that a number of factors influence building users, and their experience of one may affect their judgement of another. An important stage of BIU assessment is therefore following up on user feedback and using other measurements to determine performance problems and likely causes of low comfort ratings.

In addition to BIU assessment, there is an ever-increasing choice of instruments for collecting and measuring user feedback on environmental conditions in buildings.[2] The human or qualitative comfort rating is usually arrived at by the summation of a wide variety of perceptions and judgements to do with human memory over a given period. However, a new condition or one that causes particular concern may be subjected to short-term judgement that is not indicative of the long-term operation of the building. No single type of user feedback technique predominates: in addition to measuring instruments, evaluators may interview users, question them on psychosocial factors, such as employer–employee relations and seek feedback on other factors that influence health and morale and, therefore, environmental judgements. Moreover, there are important decisions to be made in the selection of diagnostic measuring instruments for gathering follow-up data on building performance in areas such as: indoor air quality and ventilation performance; thermal comfort and humidity; lighting and visual comfort; and noise levels and acoustic comfort. In each of these cases, the researcher must determine how to approach qualitative and quantitative measurement: data-logging over an extended period, usually in a limited number of places; or spot checks in a compressed timespan, but over a larger geographical area. A further issue is how to calibrate the instruments, in terms of their baseline settings or standard of comparison.

Applying a theory of user comfort to feedback results

Although users are not routinely consulted as part of building design, procurement and even occupancy, the above considerations indicate that both techniques of effective user input and constructive and cost-effective ways of applying feedback from users to decision-making are complex. In an effort to disaggregate and study some of the ways in which users are influenced in their assessments of the environments they occupy, a theoretical model of user comfort in buildings is being developed that permits some of the complexity of user responses and perceptions to be classified. Such classification facilitates not only future research but also the honing and refinement of feedback techniques to increase the validity of the information received and to render it more directly relevant to building decision-making (Vischer, 2005).

The model is illustrated in Figure 15.2. It distinguishes between the physical, functional and psychological comfort of building users, each of which is measured slightly differently and each of which merits consideration in differing ways or at different times during building delivery and occupancy. The model explains that in real-world construction practices, decision-makers benefit most from applying user feedback to decisions on where and how to invest in building features to balance costs with the beneficial effects on users. The concept of 'investment' as a long-term perspective on building decisions – based on the value of building features and performance to clients/users – is more tangible and useful than the older cost–benefit paradigm, in which the one-time

---

[2] See, for example, the tools developed for PROBE and now available on the following web-site: useablebuildings.co.uk

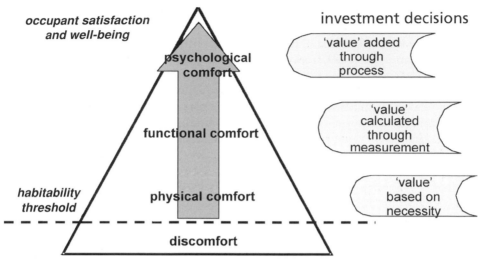

Figure 15.2   The habitability pyramid.

cost of a building feature is supposed to (but cannot, in fact) be balanced with the projected benefit to users, and thereby to the client, over time.

Investments in physical, psychological and functional comfort are evaluated differently. Decisions about physical comfort are usually basic to building habitability and depend on existing codes and standards. Investment in functional comfort is assessed according to systematically gathered user feedback on tasks and activities and the environmental requirements thereof – such as is provided, for example, by BIU assessment. Finally, investing in psychological comfort should not be ignored, as all user responses and perceptions are influenced by psychosocial factors. The principal components of psychological comfort are territory, status and control. The more users are systematically and explicitly engaged in the planning and design process and kept informed, as suggested by the BPE cyclical feedback loops, the more effective the impact on users' psychological comfort and their eventual level of comfort in the building.

Using this broad theoretical approach to user involvement generates not only appropriate tools and techniques for participation and feedback, but also a knowledge base that decision-makers can draw on to apply more of a user orientation to each stage of building decision-making. This helps to organise and classify the results of user feedback initiatives and to develop ways of making different types and levels of feedback directly relevant to building decisions.

Looking to the future

In this chapter, various ways in which a more user-oriented perspective can thrive in the context of financially driven building delivery and operation processes have been highlighted. In spite of years of effort, the two main applications of user needs research – briefing and POE – are not carried out as often or as well as they could be, and, in the case of POE, usually not carried out at all. One way of integrating these and related user feedback activities into building decision-making is to apply the BPE approach to industry-wide decision-making. A number of tools and techniques exist to capture, analyse and apply feedback from stakeholders at different stages of the construction process. In addition, a theory of user comfort, that is more complex than the common measure of simple satisfaction, can help to organise this information and provide decision-makers with a way of determining the *value* of different environmental investments. Thus the purpose of the feedback loops built into the six BPE phases is to provide opportunities and mechanisms for gathering information on which decisions about quality can be based. Quality is composed partly of generic facts about human comfort generally – human comfort in buildings in particular – and partly of information that is specific to a situation or building project. The value of gathering generic information from data banks such as design guidelines, codes and standards, research results and the like, as well as applying user surveys, focus groups, observation and other techniques is defined in terms of increased environmental quality.

How, then might one expect a building industry to change from what we have now to one focused on quality? In applying BPE, all the phases of strategic planning, briefing, design, construction, occupancy and adaptive re-use/recycling will be carried out routinely and legitimately on every project, and not just in some situations with special requirements. Financial lenders will calculate their loans and incentives for new building projects not just according to short-term returns, but also by taking quality criteria for long-term building operation into consideration. The time invested in ensuring that the feedback loops of each phase are implemented will be more than compensated by the reduced time spent later on in the life of the building correcting problems, repairing elements that do not work and adapting to new uses. Users will feel empowered instead of imposed upon: one of the spin-off effects of consulting them about their needs will be better informed users capable of and interested in participating in decisions about their environment. This will translate into fewer complaints and service calls, and a constructive partnership between users and facilities managers.

The component parts of BPE are already in existence and proven in many parts of the world. In order to see a complete implementation of the BPE approach, the process of linking them up must be set in motion by competent and experienced professionals. They must ensure that decision-makers are involved in each of the phases, and carry through on applying user feedback to building design, construction and operation. The leaders of such a process require training in group process and communication, as do all building professionals, including architects, project managers, interior designers and facility managers aspiring to bring quality to our built environment. A case study example related to this area is given in Chapter 18.

Conclusions

In view of the proliferating amount of user-needs research and initiatives that have been established in numerous countries, and the number of other books and projects they in turn refer to, we can only speculate on how much longer the commercial building industry will continue to place the needs of and our knowledge about users at the end of their list of priorities. Small gains are being made as more effort is expended on proper briefing, as POE becomes more prevalent and as a growing number of people outside the traditional closed circle of financier–developer–designer–builder are becoming involved both in supplying buildings and in examining details of their operation over their lifetimes. These include researchers, process facilitators, non-traditional architects, interior architects and managers, as well as traditional building industry professionals. Some of these initiatives are technical, some are cost-oriented, and some are humanistic – but all point in the same direction: change. It is inevitable that as more knowledge accumulates and is disseminated, and as momentum gathers to change traditional building processes and recognise the complexity of modern and future construction, that optimising the quality of the built environment which people occupy will become the major goal of construction.

16 Construction is good!

Angela Lee and Peter Barrett

Summary

Chapter 9 stressed the need to promote a positive view of construction. A core debili-
tating assumption is that the industry is unhelpfully fragmented. This chapter draws
evidence from various studies and argues that the perceived 'constraints' of construction, introduced in Chapter 1, are in fact natural characteristics of the industry. This
then leads to a different perspective, the stress of which is how to work with a highly
differentiated industry made up of many, predominantly small, organisations, rather
than seeing its fragmented nature as the problem. Findings from several studies are
used as evidence to argue that the constraint of fragmentation and the large number of
small organisations in the industry are often misconstrued as 'the problem'. Perhaps,
instead, they should be accepted as characteristics of the industry that have developed
over time as it adapts to successfully meet the challenges of its turbulent business
environment. Thus, the debate shifts from questioning the industry's 'lamentable' performance owing to its project-based characteristics, and towards the view that it is
appropriately differentiated but not well integrated. From this position, the spotlight is
turned to the need for better integration, and it is suggested that the problem lies more
at the project–company and company–company interfaces than at the project level itself. Further, the preponderance of small and medium-sized companies in the industry
means that big initiatives for big companies may be necessary, but are not sufficient.
This is highly relevant at a policy level, where there also seems to be unnecessary differentiation between the construction industry and its suppliers and specialists, and,
more conceptually, a lack of integration between the industry itself and its role in society. With this perspective in mind, the reasons behind why the industry is thought to
have failed to maximise its performance so far are pursued.

The nature of the construction industry

The construction industry is Europe's largest industrial employer, representing 7.2%
of the continent's total employment and 9.9% GDP (FIEC, 2003; see Table 16.1).

It is one of the UK's largest industrial sectors, providing employment for 8% of
the UK working population (DTI, 2003a). With its predominance in the economy, the
construction industry, however, represents one of the most complex and dynamic industrial environments. It relies heavily on skilled manual labour that is supported by an

164

Table 16.1 Number of construction employees and production by country.

Country	Number employed (× 1000)	Production (€bn)
Europe		
Austria	258	28
Belgium	230	28
Germany	2418	213
Denmark	162	20
Spain	1825	103
Greece	320	14
France	1512	118
Finland	146	17
Great Britain	1629	139
Italy	1748	107
Ireland	186	20
Luxembourg	30	1
Netherlands	483	49
Portugal	618	24
Sweden	215	24
Total	11 780	905
Other		
Japan	6230	626
USA	6544	880

Source: FIEC, 2003.

interconnected management and design input, which is often highly 'fragmented' right up to the point of delivery (Mohsini & Davidson, 1992). A large and complex project will involve many design, construction and supplier organisations, whose sporadic involvement will change throughout the course of the project (for examples, see Carty, 1995). The organisations will be both large and small, and although they have usually never met before, they are expected to work together effectively and efficiently through-out the duration of the project (Kagioglou et al., 1998; Lee et al., 2000). Complicating this situation yet further, the vast majority of design and construction activities are subcon-tracted, which renders collaborative and integrated working extremely problematic. In addition, design and construction practitioners typically find themselves working on several projects at the same time. According to Mullins' (1999) generic and rather simplistic prescription, the success of a project relies heavily on having clearly defined objectives and well-defined tasks. But these are not always feasible in construction where the client's objectives themselves often only crystallise over time.

Moreover, the entire construction labour market is founded on widespread self-employment (Briscoe et al., 2000). The scale of small organisation activity in the UK construction industry is considerable, with, in 2002, 99.3% of UK construction organi-sations having 1–79 staff and employing 65.4% of the total construction workforce (DTI, 2003a; see Figure 16.1). The predominance of micro enterprises in the UK construction

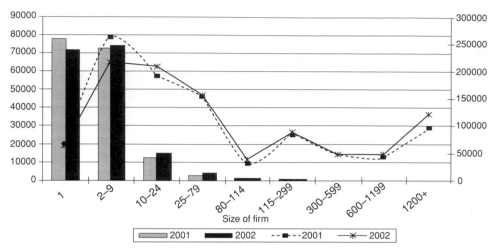

Figure 16.1   Number of UK construction organisations and employment figures in 2001 and 2002 (DTI, 2003a).

industry, employing between 1 and 9 persons,[1] over large-sized organisations (see Figure 16.1) may be attributed to the fact that large contracts require specialist work and the specialist contractors are predominantly self-employed and, where necessary, employ a few additional hands (Abdel-Razek & McCaffer, 1987; Gale & Fellows, 1990). According to Langford & Male (1992), larger organisations generally resort to a greater use of subcontractors (micro and small enterprises) in a bid to reduce the overhead burden of tax, National Insurance contributions and working capital needs. This opposes Pearce's (2003) earlier proposition that small firms 'inhibit the capture of economies of scale', or at least factors in the likely incidence of diseconomies of scale. Thus, any overall performance improvement of the industry through innovation will be significantly influenced by small construction organisations, of various trades, given that they make up the majority of the industry.

The predominance of SMEs in construction is not unique as is often thought. The agriculture, hunting, forestry and fishing sectors follow a similar pattern to that of construction. For example, Figure 16.2 illustrates the employment of UK enterprises by size of organisation. It is evident that there is an abundance of micro enterprises, but due to the scale of the chart the number of large enterprises is not clear. Hence, it is appropriate to also view the total number of persons employed in context. The manufacturing sector is seemingly a mirror to that of construction – more persons are employed in large enterprises than micro and small enterprises. Only the 'wholesale and retail trade/repairs', and 'real estate, renting and business activities' sectors are relatively balanced in terms of the number employed in micro and large enterprises. However, only the construction industry, with the exception of manufacturing in some instances, operates within a project-based environment with temporary project teams

---

[1] This paper will adopt the European Commission's definition of 'small and medium enterprise' (SME); whereby micro enterprises represent 0–9 employees, small enterprises represent 10–49 employees and medium enterprises represent 50–249 employees (with the exception of agriculture, hunting, forestry and fishing organisations; European Commission, 2003).

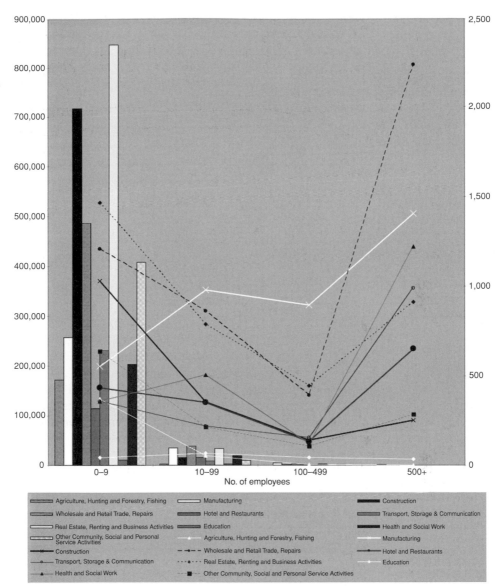

Figure 16.2   Number of UK enterprises and employment figures by industrial sector in 2001 (SME statistics, 2002).

of various disciplines – hence, its very fluid nature. The structure of the industry is arguably a function of the work it is called upon to do. In some ways it seems closest to 'other community, social and personal services'!

Performance of the construction industry

The UK's Department of Trade and Industry (DTI) has investigated the relative performance of various industrial sectors. Figure 16.3 illustrates the gross value added

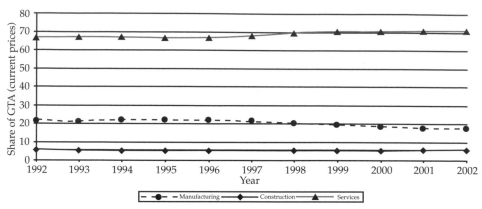

Figure 16.3   GVA contributed by UK manufacturing, construction and service sectors (DTI, 2003b).

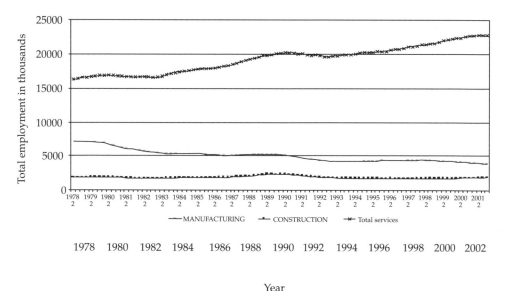

Figure 16.4   UK employment trends of manufacturing and construction and service sectors (DTI, 2003b).

(GVA) by construction, manufacturing and service sectors. This shows that, although construction contributes less than manufacturing in general, its contribution is holding steady against a declining profile for manufacturing and a rising trend for the service industry. To an extent, these changes reflect the changes in the volume of activity going on in the various sectors, and in particular, the significant growth of services. It is interesting to look at relative employment trends in manufacturing and construction over the longer term (two decades) as shown in Figure 16.4. From this it is clear that employment in manufacturing has and continues to decline, but the picture for construction is, in simple terms, cyclical, indicating considerable flexibility in the face of changing

Figure 16.5 Relative GVA of UK manufacturing, construction and service sectors (DTI, 2003b) indexed from 1976.

demand. This does not, however, indicate whether the given sectors are working more efficiently. The declining numbers of those employed in manufacturing could have resulted in much greater GVA. To throw light on this, Figure 16.5 tracks the relative trends in GVA over the last 25 years. From this it can clearly be seen that construction has actually improved its contribution much more steeply than manufacturing, albeit less so than services, with the latter fuelled by absolute growth in activities.

This is not a picture of a failing industry, or of 'lamentable' performance. It indicates an industry responding robustly to highly turbulent conditions and making a valuable and increasing contribution to the economy. The 'fragmented' structure and predominance of SMEs in its project-based environment should indeed be accepted as positive characteristics of the industry, which have developed over time as it adapts to successfully meet the challenges of its turbulent business environment.

Fragmentation versus differentiation

The aspiration to enhance construction performance through innovation has often been stunted by the general assumption that the intrinsic characteristics of the construction industry – such as fragmentation, 'boom-and-bust' market cycles, use of relatively low technology and antagonistic procurement policies – inhibit innovation (Ball, 1988; Brouseau & Rallet, 1995; Powell, 1995; CERF, 2000; Gann, 2000; Fairclough, 2002). Although it is acknowledged that construction organisations have always demonstrated an ability to innovate (for example, see Slaughter, 1998), there is a negative connotation

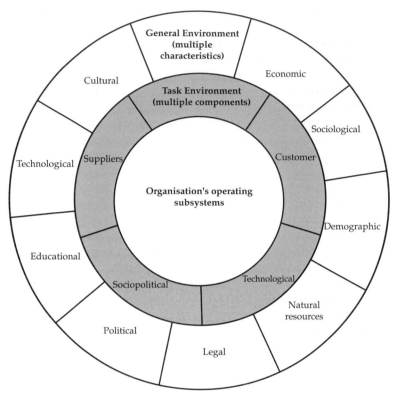

Figure 16.6   Relationship of general and task environments to the organisation (adapted from Kast, 1985).

to the choice of the adjective 'fragmented' to describe the structure of the industry. According to the *Oxford English Dictionary* (1993), the term 'fragmented' denotes
'broken or separated into fragments; disjointed', and thus 'fragment' signifies 'detached
or incomplete part'. Although the construction industry is composed of various specialist organisations, is it justified to describe the sector as disjointed and incomplete?
The nature of the industry is that of various disciplines – the parts – which temporarily
form together as the design and construction project team and then reconfigure for the
next project.

This chapter proposes that by recasting 'fragmentation' as 'differentiation', the debate
shifts from the structure of the industry as a 'problem' to it being an inevitable, indeed,
appropriate response that provides 'requisite variety' to meet the complex and dynamic
nature of the construction environment (Beer, 1985). The design and construction of
buildings is carried out by organisations that act more like coalitions than teams (Fellows et al., 1983). The manner in which the coalitions reach compromises, or better yet,
synergies, between their different goals is a major factor in determining their effectiveness. The level to which the added value contributed by each of the parties involved
is mutually understood, significantly determines the potential success of a building
project during the process of initiation, briefing, design, engineering, construction and

use. Construction organisations are sub-systems of a broader spectrum comprising both the project/task and the general environments in which they operate (Figure 16.6). The project or task environment, as discussed earlier, has a direct impact on the goals and values, structure, technology, human relationships, managerial processes, etc. within the organisation. There is evidence that project imperatives get higher attention within construction companies (Sexton & Barrett, 2003a, b). These influences may very well impact differentially creating conflict in the organisation's own operating sub-systems. In addition, organisations are becoming increasingly subject to external forces – the general environment (political, legal, cultural, sociological, etc.) – and this is clearly evident in today's marketplace. For example, automobile companies are hearing more directly not only from customers concerning safety and performance, but also from environmentalists concerning air pollution; the hospital is being called upon to expand its boundaries and to deal with the total health care needs of people in the community. On all fronts, including the construction industry, organisations are facing a more heterogeneous and uncertain environment. Thus, differentiation as stressed in this chapter, refers to the different goals, time-horizons and working practices of different parts of the organisation that operate within its project units (adapted from Robbins, 1979, Lawrence & Lorsch, 1967, and French & Bell, 1984). Thus, the project focus helps to explain the nature of the construction industry, and where large construction organisations are concerned it highlights the source of diseconomies of scale.

Differentiating by product instead of specialism is a possible solution, but it simply recasts the problem (Burns & Stalker, 1961; Kast & Rosenzweig, 1973). What remains, irrespective of this choice, is the importance of integrating mechanisms so that a coherent provision to customers is delivered. It may well be found in the large-project (or high-volume, small-project) domain that differentiating by product is feasible and desirable. In the small-project domain, it may remain more sensible for functional specialists to build sufficient volume of business through involvement in multiple projects in order to maintain the critical mass to support the specialism in question. This is a common issue in matrix organisational design (Galbraith, 1977) and there is a real danger that the specialist knowledge underpinning much of the work of construction could be lost if these 'experts' became dispersed across a large number of product-orientated companies. It may be that the historical arrangements have emerged, not as anachronistic legacies, but because, through some Darwinian process, they do actually meet the needs confronting the industry (Barrett, 2002), as Figure 16.3 would suggest. That is, to flexibly and constantly reorganise to achieve complex problem-solving in highly unstable economic situations.

Differentiation thus integration

This then leads to a stress on how to work *with* a highly differentiated industry made up of many, predominantly small, organisations, rather than seeing it as the problem: the industry is highly differentiated, but not well integrated.

Integration is defined in a general sense as 'the quality state of collaboration that exists among departments that are required to achieve unity of effort by the demands of the environment' (Robbins, 1979). Different environments require different amounts of integration, and integration between different work units. Thus, the environment determines the requirements for differentiation within and among the work units, and integration within and among the work units (French & Bell, 1984). Differentiation and integration represent opposing forces; the key is to appropriately match the two. An organisation needs to differentiate to deal with the specific problems and tasks it faces; but the more differentiation, the more energised mechanisms are needed to get people to integrate and work as a cohesive coalition towards organisation and project goals (see Figure 16.6).

The imperative to effectively integrate highlights the importance of broadening the conception of construction to include suppliers and specialist contractors, and ought to extend immediately to facilities management (FM). This can be taken much further to question the very purpose of construction. Construction is a means to improve the built environment so that it can better support the quality of life and competitiveness of society. Construction is not an end in itself (Barrett, 2003). Thus, customer, cultural, educational, legal, political, etc. issues must also be considered. Moreover, there is another equally important side to this equation. Instead of asking how well the small business serves the labour market, there is the opposite question of how well the small business itself is served by the labour market. This has been seriously overlooked at policy level until recently, at least in the UK. Policy forums such as CRISP (Construction Research and Innovation Strategy Panel) took an inward-looking construction industry perspective and tried to identify with major industry players what they needed, which led to a fairly sterile debate.[2] In addition, there is an understandable desire at policy level for clear prescriptions of best practice for the industry to take up, and these are looked for amongst large organisations with small organisations seen as the problematic laggards. Given the preponderance of small organisations in construction, with distinctive innovation characteristics and capacities (Sexton & Barrett, 2003a, b), this quest with major players for simple answers to complex, dynamic problems has arguably at times obstructed rather than supported progress.

The scale of SME construction activity in the UK should be duly recognised. It is no longer possible to discuss strategic policy (economic, innovation, etc.) without recognising the positive role SMEs play in the sector. Policy must be appropriate to SMEs and appropriate for those (large organisations) whose environment is influenced by the activities of SMEs. This means that policy towards the sector has to examine the implications of SMEs within a wider framework than has been the case in the past. There has been a lack of integration between the practical realities of the industry and the scope and orientation of public policy for research and innovation, and this should be addressed.

Conclusion and the way forward

This chapter sought to address the problem of 'what needs to change' for (appropriate) innovation in construction to be achieved, and in particular, should the industry be less fragmented?

---

[2] This observation is based on direct involvement in the first two generations of CRISP.

The construction industry operates in a project environment, where there is a high degree of 'fragmentation' and interdependency between organisations. Yet the industry, through the products that it creates, its size and ability to create employment, has the potential to influence an economy's gross domestic product (GDP) more than any other service industry. It is at least as tenable that the industry has successfully evolved and adapted compared to many other industries that have suffered catastrophic failure. It is quite probably appropriately 'differentiated' so that it can sustain the specialist areas of knowledge needed whilst flexibly bringing them together around temporary projects, constantly re-organising to achieve complex problem-solving in highly unstable economic situations. The construction industry has unusual characteristics, which are related to its structure, production process, physical characteristics and composition. Those characteristics go a long way towards explaining methods of production, organisation, price determination, payment methods, financial decision and control, and an industrial structure unlike those met in other sectors. Thus, this chapter proposes that the expression 'differentiated' better describes the characteristics of the industry rather than the infamous 'fragmented' term that is commonly used to describe the structure of the sector.

In summary, it is suggested that organisations, especially small construction organisations, are well integrated and highly responsive to project needs, but that this is often at the expense of enduring company-based improvements. This, in turn, undermines the integration needed to support company-to-company improvements, especially given the lack of stability in the construction economy. Value network innovations are difficult in practice, but are also placed in a policy vacuum by an over-emphasis on a tight conception of the construction industry itself, in which major organisations are perhaps too glibly taken as the template for all organisations to follow. So, the industry is highly differentiated, but at a project level the integration effort is generally kept in clear focus. It could be said that 'single loop learning' (Agryris & Schon, 1978) – that is, pragmatic problem-solving – to 'do things right' on the ground is alive and well. The same cannot be said for longer term, company-based innovation or the policy framework within which it is placed. Here 'double loop learning' is severely limited by the turbulence of the industry's workload and the limited resources of SMEs, so that progressively moving towards 'doing the right things' is hard to sustain.

A pragmatic critical-realism approach is suggested in which the real world is accepted as an inconveniently complex and dynamic object to study and work with. Indeed in Chapter 17 the argument is made that in developing countries some of these industry 'peculiarities' can be seen as positive assets. So, the focus is not directly on events, but rather on the identification of candidate generative and contingent mechanisms, tested for practical adequacy in the real world, hence the pragmatic emphasis (Johnson & Duberley, 2000). The first step to achieve this would be to embark on creating an integrated, co-ordinated and research-based policy framework for construction. Once efforts are harmonised, the balance between differentiation and integration for a more innovative construction industry can be addressed. More importantly, however, initiatives should embed the reality that the construction industry is largely composed of SMEs, and is differentiated rather than fragmented. From this a strong focus should be promoted to discover the key generative mechanisms that can underpin effective integration.

17 A wider view: revaluing construction in developing countries

George Ofori

Summary

There are attempts in many countries to undertake strategic and structural reform of the construction industry. In many of these exercises, the construction industry is being revalued. The key objectives are to improve the performance of the industries, prepare them to face the perceived challenges in the future and thus to enhance the image of the industries in these countries.

While much of the material in this book draws from experiences in 'developed' economies, this chapter considers the revaluing of the construction industries in developing countries, based on a review of the literature. The chapter starts by defining 'developing country', but pointing out that the countries referred to as such are not homogeneous. It discusses efforts to develop the industries of these countries and goes on to consider the revaluing of the industries, looking at two trends which need to be addressed: globalisation and private-sector participation in investments in construction. In addition, some of the features of construction and their implications for the revaluing process are addressed. Finally, a number of proposals for action are made. It is suggested that the construction industry should take a long-term, strategic orientation, develop country-specific recommendations and champion the development of the industry.

Introduction

This chapter is a contribution to a debate on the revaluing of construction initiated by the International Council for Research and Innovation in Building and Construction (CIB) in 2001 (Courtney & Winch, 2003). Winch et al. (2003) consider the revaluing of construction to have three aspects:

(1) Generating better understanding of how constructed assets add value for their clients and users (they consider four aspects of this value: spatial quality; indoor environmental quality; symbolism; and financial value as a capital asset for exploitation or sale)

(2)   Developing a more effective capture of the value generated through the project life-cycle in terms of profits and learning
(3)   As a result of the two previous points, a revaluation of the image of the industry and the way it is perceived by those outside it

In this chapter, in discussing the revaluing of the construction industry from the perspective of developing countries, focus is put on the development of these industries in order to improve their performance and prepare them to face their future tasks and challenges for the benefit of these nations and their people.

The aim of this chapter is to consider the revaluing of the construction industry in the developing countries. The objectives of the chapter, which is based on a review of the literature, are to:

- Highlight the importance of the construction industry in developing countries.
- Discuss the efforts that have been made to improve the performance of the industry in these countries.
- Consider current relevant trends that have implications for these industries.
- Propose possible initiatives for improving the performance of the construction industry in developing countries in order to enhance their public image and their value to society.

It is necessary to define the group of countries to which this chapter relates. There is a continuing debate on the most appropriate definition of the expression 'developing country', but any detailed consideration is beyond the scope of this chapter. This chapter adopts the definition of the major international financial institutions such as the World Bank and the International Monetary Fund, who define a 'developing country' by considering the level of per capita national income. Currently, the World Bank defines a 'low-income' country as one with a per capita gross national income (GNI) of US$765 or less in 2003; and a 'middle-income' country as one with a GNI of US$9385 (World Bank, 2004a). The group of low-income and middle-income countries are collectively referred to as 'developing countries'.

It must be noted that the large number of countries classified as 'developing' do not constitute a homogeneous group. This is evident from the range of incomes, and, indeed, the World Bank further sub-divides the group of middle income countries into 'lower-middle-income' and 'upper-middle-income' at a GNI per capita of US$3035 in 2003 (World Bank, 2004a). The 'developing countries' differ in terms of size, current level of income, economic growth and future prospects. These can be seen from the data on 'selected indicators of development' presented in the World Bank's annual World Development Reports (see, for example: World Bank, 2004a, pp. 256–264). Table 17.1 shows a few indicators of the economies of 12 developing countries. The data illustrate the differences among the nations in terms of population, national income, income per capita, inflow of foreign direct investment and total external debt. The latter two figures give an indication of the economy's future growth prospects.

It is also pertinent to note that the developing countries differ with regard to their construction industries, including: size of industry; resource endowments; level of technologies being applied; practices and procedures; and operating environment. For

Table 17.1 Some development indicators of selected developing countries.

Country	Population (million)	GNI (US$bn)	GNI per capita (US$)	Foreign direct investment (US$m)	External debt (US$m)
Angola	13.5	10	740	1312	10 134
Bangladesh	138.1	55	400	47	17 037
Burundi	7.2	1	100	0	1204
China	1288.4	1417	1100	49 308	168 255
Ethiopia	68.6	6	90	75	6523
Honduras	7.0	7	970	143	5395
India	1064.4	568	530	3030	104 429
Papua New Guinea	5.5	3	510	50	2485
Paraguay	5.6	6	1100	−22	2967
Philippines	81.5	88	1080	1111	59 342
Zambia	10.4	4	380	197	5969

Source: World Bank (2004b)

convenience, the developing countries are considered in this chapter as a group but where relevant, the differences are highlighted.

The literature highlights the importance of the construction industry as a sector of the economy, and draws attention to its size, in terms of its contribution to overall gross national income, employment (Hillebrandt, 2000) and its potential backward and forward linkage effects with several other sectors (World Bank, 1984). The industry also contributes to capital formation, and hence its products represent a significant proportion of each nation's savings. Considering this role as the producer of the nation's physical infrastructure and other productive assets, the industry is important in national development.

In the developing countries, more than anywhere else, the construction industry has a critical role to play in the economies. Thus, although the construction industry is a national asset, it tends to be seen as a 'spender of investment'. However, in most developing countries, the construction industry has failed to play its expected role in providing the basis for socioeconomic development, as well as that for securing improvements in the living conditions of the citizenry. Thus, backlogs of unfulfilled construction needs continue to build up, which is further eroding the basis for national economic growth. The industry has not been able to be 'the engine of growth' that it is widely considered to be (see, for example: World Bank, 1984; Wells, 1986) by stimulating activities in other sectors of the economy.

Development of construction industries

Recent major industry reviews

Since the mid-1990s, several countries at different levels of development have undertaken comprehensive reviews of their construction industries and formulated long-term plans for improving their performance. These countries, which are mainly 'developed' nations, include: Australia (Australian Procurement and Construction Council, 1997); Hong Kong (Construction Industry Review Committee, 2001); Singapore (Construction 21 Steering Committee, 1999); and the UK (Latham, 1994; Egan, 1998). The reviews were undertaken because of perceived weaknesses of the industries, and internal and external threats and challenges. In particular, there was evidence that clients, users and other stakeholders were dissatisfied with the performance of the industry (see, for example: Nkado & Mbachu, 2002). There were also expectations of greater complexity of future constructed items in increasingly sophisticated economies, technological and social change as well as globalisation leading to competitive pressures at home and abroad.

The industry review initiative in Singapore is a good example of such efforts. In the context of this chapter, it is appropriate to discuss the recent experience of Singapore as the country has progressed from a typical developing country to a high-income nation (in terms of GNI per capita) in about 30 years. A full account of the construction industry development effort in Singapore can be found in Ofori (1993b). The review of the construction industry in Singapore, called the 'Construction 21 (C21) exercise', was one in a series of studies on various sectors of the Singapore economy towards the turn of the century, including *Manpower 21* (Singapore Manpower 21 Committee, 1999). The C21 study was spearheaded by the Ministry of Manpower and Ministry of National Development in a bid to improve the productivity of the industry by a radical restructuring of its processes, procedures and practices (Construction 21 Steering Committee, 1999). A steering committee, with representatives from the main segments of the construction industry, had the following terms of reference:

- To arrive at a vision and role for the construction industry in the twenty-first century
- To examine the current status of the construction industry with regard to techniques, manpower, management practices and others, and benchmark it against the best standards in the world
- To set concrete targets for the construction industry and its workforce in Singapore by taking into account the gaps between current reality and the intended vision
- To recommend strategies to meet the targets and move the construction industry and workforce towards the intended vision

The C21 Steering Committee formulated a vision for Singapore's construction industry for the twenty-first century: 'To Be a World Class Builder in the Knowledge Age'. Under the vision, C21 proposes to transform Singapore's construction industry from 'the 3Ds' to 'the 3Ps', i.e. from an industry which is 'dirty, dangerous and demanding' to one which is 'professional, productive and progressive'. The report highlighted the following desired outcomes for the Singapore construction industry:

- A professional, productive and progressive industry
- A knowledgeable workforce
- Superior capabilities through synergistic partnerships
- Integrated process for high buildability
- Contributor to wealth through cost competitiveness
- Construction expertise as an export industry

Under the vision, six strategic thrusts were identified:

- Enhancing the professionalism of the industry
- Raising the skills level
- Improving industry practices and techniques
- Adopting an integrated approach to construction
- Developing an external wing
- Developing a collective championing effort for the construction industry

The implementation period is 2000–2010, with the Building and Construction Authority (BCA) identified as the 'champion agency' and the Construction Industry Joint Council (CIJC) also assigned some responsibilities. The latter is an umbrella group for the professional institutions and trade associations in construction. Significant achievements have been made in: information technology applications; integrated procurement approaches, such as design and build; formulation of a national quality and productivity framework; and promotion of the export of services. However, problems still remain, particularly in the areas of sustainable demand for construction, human resource issues as well as health and safety. This indicates that construction industry development is a long-term process. In the next section, studies on the construction industries in developing countries are discussed chronologically.

Studies of construction in developing countries

The construction industries of developing countries have been studied since the 1950s (Ofori, 2001). The first of these took place over several decades, inspired by efforts to improve the poor housing conditions in these countries, and the realisation that the industries which must develop the much needed residential units needed to be built up. For example, the United Nations appointed an Expert Group on Housing, which, as part of its recommendations (Department of Economic and Social Affairs, 1962), proposed a programme for developing the construction industries in arguably the first published comprehensive study on the construction industries of these countries. The Economic Commission for Africa (1965) also examined how the housing conditions in the poorer countries could be improved, and again recognised that the construction industries had to be strengthened if they were to be able to meet the huge volumes of housing needs. The focus on housing led to the First United Nations Conference on Human Settlements in Vancouver, Canada in 1976 which, in turn, led to the establishment of the United Nations Centre for Human Settlements (UNCHS) (now the United Nations Human Settlements Programme (UNHSP)), based in Nairobi, Kenya.

In the late 1960s and 1970s, the key concern was with the need to enhance the capacity and capabilities of the construction industry to enable it to establish the physical basis for national socioeconomic development (see, for example: Turin, 1973; Ofori, 1993a). Among those who worked on the industry under this agenda were the University College Environmental Research Group in London. Turin (1969, 1973), who led this group for several years, hypothesised the role of construction in the process of socioeconomic development by examining the relationships between: value added in construction and GDP; capital formation in construction and gross domestic capital formation; and employment in construction and the total workforce. On the basis of his findings, Turin proposed a strategy for improving the construction industries. Other researchers have tried to build upon the foundations established by Turin and others, and several of them have tested and largely confirmed the relationships between construction and the economy which he first highlighted (see a review in Ofori, 1993a, b).

There have been four main international agencies involved in the effort to develop the construction industries, the:

- International Labour Office (ILO) – which has a construction sector adviser. The ILO had a construction management programme from the late 1970s to the mid-1990s, which covered countries in Africa and Asia (see, for example: Miles & Neale, 1991). Among its greatest contributions was work on labour-based construction (which continues under the 'ILO Assist' programme based in Harare, Zimbabwe and Bangkok in Thailand); and also work on contractor development which led to a definitive set of guidelines for such initiatives (ILO, 1987), and continues under the 'Improve Your Business' programme.
- UNHSP – which has published several important reports on policies, procedures and practices (see, for example: UNCHS, 1991, 1995).
- United Nations Industrial Development Organisation (UNIDO) – which produced some key works on the materials production industries and technology transfer (see, for example: UNIDO, 1969).
- World Bank – which has also produced many reports on the development of the industries (see, for example: World Bank, 1984; Kirmani, 1988; Kirmani & Baum, 1992). The World Bank has launched several industry development programmes in different countries, including training, equipment hire-purchase and building materials production. In the procurement of the projects for which it provides funding support, the World Bank continues to provide a 7.5% tendering preference for contractors from developing countries.

Indeed, reviews of national construction industries and their performance have been undertaken in the developing countries. An example worth highlighting is the study on Tanzania undertaken by a group of international experts in the mid-1970s (Ministry of Works of Tanzania, 1977). Among other actions, this comprehensive report led to the establishment of the National Construction Council (NCC) to manage the development of the industry. Similar agencies that have been formed elsewhere include: the Construction Industry Development Board in Singapore (formed in 1984, which became the Building and Construction Authority (BCA) in 1999); the CIDB in Malaysia (set up in 1994); and the CIDB in South Africa (formed in 2002).

In 1997, the CIB Task Group 29 (TG29) on Construction in Developing Countries was established. TG29 became a working commission (W107) in 2002. The aim of W107 is to help build the body of knowledge on the construction industries in developing countries, and to develop and disseminate the ideas, tools and techniques which can be applied to improve the performance and prospects of these industries. It is making a contribution to the research and practical effort on industry development. The areas on which W107 is currently focusing include:

- Poverty alleviation through construction
- Sustainable construction in the context of developing countries
- Safety and health in construction, including community health, especially the relationship between construction and the propagation of HIV/AIDS
- Implications of privatisation for construction enterprises and practitioners in developing countries
- Merits of a central agency for managing construction industry development in developing countries, and the potential of regional groupings of such national agencies
- Implications of globalisation for local enterprises in developing countries
- Application of information technology in construction in developing countries
- Appropriate construction management and economics techniques for developing countries

Studies by individual researchers continue, although the volume is smaller compared to the levels in the 1980s and early 1990s. For example, Ofori (2000b) considered some of the challenges facing the construction industries in developing countries, focusing on construction industry development, globalisation, culture and the environment. He drew lessons from the recent experience of other countries at all levels of development, especially Singapore, to highlight possible ways by which progress could be made in the developing countries. He then presented a research agenda for improving the performance of the construction industries of the developing countries in the light of their resource constraints and administrative weaknesses.

Attempts have been made by other researchers to revalue aspects of construction in developing countries. These include:

- Assessing the readiness of the industries to cope with the knowledge economy (Ofori, 2002; Ebohon, 2003)
- Discussing the impact of globalisation on the industries (Ofori, 2000b; Muhegi & Malongo, 2004)
- Considering how the industries in developing countries can use information technology most effectively
- Evaluating the relevance of various management techniques to the industries (Shakantu et al., 2002)

Results of development efforts

What are the results of the efforts to develop the construction industries in developing countries? Ofori (2000a, 2001) notes that the construction industries in countries at all levels of development face problems and challenges. However, in the developing nations, these difficulties and challenges are present alongside a general situation of socioeconomic stress, chronic resource shortages, institutional weaknesses and, as a result of these features and circumstances, a general inability to deal with the key issues. He also observed that there was evidence that the problems of these industries had become greater in extent and severity in recent years. Some of these problems of the industries include:

- Poor cost, time and quality performance
- Lack of work opportunities
- Poor level of professionalism and entrepreneurship
- Obsolescence of some statutes and codes
- Ineffectiveness of implementation of existing statutes and codes
- Bureaucracy in formal procedures relating to project planning and administration.

Ofori (2000a, 2001) observes that the performance of the construction industries in the developing countries (with regard to the fulfilment of clients' stated objectives) has not improved despite more than a generation of research and recommendations on ways and means of improving the industries; and on efforts in many countries, in some cases, quite concerted. In this group of countries, as elsewhere, the construction industry is being compared with other sectors of the economy and being found to fall short of the mark in many respects, including productivity, quality and safety. Clients have also expressed dissatisfaction with the performance of the industries in terms of time, cost and quality, and the procedures and practices they adopt (see, for example: Ofori, 1994). In particular, serious considerations regarding environmental, safety and health issues are relatively new in these industries. This manifestly poor performance is further eroding the industry's image in these countries. For example, in his speech to open the conference of TG29 in Gaborone, Botswana in 2000, the Minister for Finance highlighted the assistance programmes offered by the Government for local construction enterprises in that country, including a funding scheme, a training programme and a contract reservation policy, and then expressed the public client's dissatisfaction with the performance of the local contractors. In Malaysia, a number of failures on large construction projects (including cracks in the columns supporting a flyover on a highway, leading to closures and heavy traffic disruption; a general hospital which had to be closed a few months after it had opened because defects in the air-conditioning system led to a serious fungal outbreak; and a major multi-storey public office building which is 7 years behind schedule) has been the subject of national debate (Ahmad, 2004; *The Straits Times*, 2004).

Revaluing construction

Features of construction industry from perspective of developing countries

The construction industry contains many features that are often considered to be 'problems' in the literature on construction (see, for example: Nam & Tatum, 1988). Dubois & Gadde (2002) and Koskela (2003) are among authors who question this characterisation. Koskela (2003) referred to these features as 'peculiarities' and used them as a framework to consider the revaluing of construction. In the context of the developing countries, these inherent features can be looked at as 'opportunities' and 'strengths' in the context of the developing countries. Some of these features will now be considered.

The industry's location specificity (see, for example: Hillebrandt, 2000) is considered to pose managerial problems – in particular, those relating to logistics and the management of risks emanating from operating in potentially unfamiliar physical, business and legal environments. For developing countries, this location specificity means that no matter how international the group of participants in a construction project is, they must 'act local' in the sense that the item must be built on location, and in the legal, cultural and resource context of the site. This provides opportunities for job creation, in all parts of the country, for firms and individuals both in construction and allied industries. There is also the wider impact on the economy from the construction activity and the constructed product. Moreover, there are opportunities for the transfer of technology and enhancement of skills for local firms and practitioners. However, in the developing countries, the potential of construction for creating jobs (owing to its relatively labour-intensive nature) and for disseminating new technology throughout the nation (Edmonds & Miles, 1984) has not been realised. In fact, some studies have found that the industry uses less labour on a proportional basis in these countries than in the industrialised ones (see, for example: Turin, 1973). However, Chapter 14 highlights the absolute global predominance (three-quarters) of the construction labour market rooted in developing countries.

That construction projects involve organisations which participate in the project at different stages and who may never work again is considered a major weakness. It is seen as hindering innovation owing to the short-term, project-based nature of construction activity. Moreover, it is also considered to be having an adverse impact on project and team performance as a result of difficulties in securing the integration of the contribution of the different participants. Conversely though, the temporary organisations offer opportunities to different organisations to take part in construction projects. In particular, they provide possibilities for firms in all parts of the country to participate in projects, and enable the best grouping to be assembled for each project. Firms in developing countries can utilise the features of their culture to develop longer term strategic alliances, and secure the integration of the contributions of project participants.

The large size and indivisible nature of the investment in construction is also considered to be a 'problem'. In the developing countries, the banking systems are also relatively undeveloped. Thus, there is no ready availability of financial assistance for investment in construction. However, individuals are able to find practical ways to organise funding for housing and community development projects. In Ghana, much of the latter comes from donations and contributions by the members of the community, raised through the organisational efforts of elected or appointed representatives. This provides an opportunity for the construction industry to develop systems that enable it to work as a force for social cohesion in such circumstances. The large investment makes it necessary for the feasibility of projects to be carefully evaluated.

The need for regulation and control in construction is often seen as an inhibitor of technology development and innovation in construction (Nam & Tatum, 1988). However, in developing countries, this provides the opportunity for the government to manage the industry and improve its capability to meet the requirements stated in the regulations. It is heartening that in some of these nations, the governments, in their role as regulators, are also acting to deal with inappropriate practices and behaviour by introducing legislation and sanctions. This is important in countries where the statutes and codes require development. Among the latest areas where governments are beginning to impose regulations are those of health, safety and the environment.

However, it is important to find ways and means of effectively implementing existing regulations.

There are two current trends that need to be addressed in any attempt to revalue the construction industry in developing countries, those of globalisation and private-sector participation in government projects. The next section will now consider these two trends and then turn to the issue of the cultural fit of the practices used.

Globalisation

There is a continuing debate in all countries, especially the developing nations, on the merits and demerits of globalisation to these countries. It is sufficient here to say that developing countries must endeavour to harness the advantages offered by this phenomenon. The World Bank (2004a) advises these countries to improve their investment environments in order to position themselves appropriately. It is clear that globalisation is of direct relevance to construction in the developing countries. This is because, as has been pointed out by many authors over several years (such as Turin, 1973 and Drewer, 2001), many of the built items that the nations require for their socioeconomic development are beyond the capacity and capability of their industries to undertake, owing to the size, novelty and complexity of those projects. Therefore, the developing countries must import some construction services. Table 17.2 presents a summary of some of the advantages and disadvantages of globalisation to the construction industries of developing countries.

Reviewing developments in the construction industry in several Asian countries in the 1990s, Raftery et al. (1998) identified three trends:

- A greater extent of private-sector participation in major infrastructure projects
- Increasing vertical integration in the packaging of construction projects, which are growing larger
- Increased foreign participation in the construction industries of most of the countries, almost all of which are developing

They attributed these trends to 'the globalisation and deregulation of markets necessitated by fiscal technological and managerial constraints' (Raftery et al., 1998, p. 729). It is estimated that foreign contractors and consultants have about a 70% share of the construction market in the Southern African region (Ofori, 2001). The firms in developing countries must develop an appropriate response to this situation. Muhegi & Malongo (2004) estimate that foreign contractors, which constitute 4% of registered contracting firms in Tanzania, undertake more than 80% (by value) of the construction projects in that country.

It must be noted that, in globalisation, some of the exporting construction enterprises are in fact *from* developing countries, and these firms operate in countries at all levels of development. A study by Ofori (2003) of the top 30 international construction firms in the annual surveys of the top 225 international contractors in *Engineering News Record* (ENR) during 1990–1999 revealed that a total of 58 contractors were listed at least once among the top 30 contractors during the period. US contractors had the highest number

Table 17.2   Advantages and disadvantages of globalisation to construction industries in developing countries.

Advantages	Disadvantages
Involvement of international finance owing to the liberalisation of controls, and freer movement of capital, makes possible the implementation of several projects, such as those of major infrastructure.	Local construction firms have no funds or expertise to participate in the sponsorship of privatised projects.
Direct foreign investment in projects leads to increase in construction demand, creating work opportunities for local firms.	Local construction companies lack the technical and managerial capability to undertake most of the foreign-funded projects. Many of the clients of these projects engage firms they are familiar with in their home countries.
Competition among foreign firms lowers the costs of construction projects to developing countries.	It is possible that local firms will be deprived of the opportunity to grow as most of the significant projects are undertaken by foreign firms (Hillebrandt, 1999).
Presence of international firms offers scope for technology transfer, the development of local firms and upgrading of the industry. The large number of such firms also means that technology transfer can be a tool for competition.	Foreign construction firms may pay lip service to technology transfer (Carillo, 1994) or take measures to avoid it. Moreover, local companies may not be in a position to benefit from technology transfer, or to subsequently utilise the acquired expertise (Ofori, 1996).

Source: Ofori (2000b)

listed (14, or 24%), followed by Japan (9, or 16%), France (8, or 14%), UK (6, or 10%) and Germany (5, or 9%). Significantly, four of the contractors were from Asian countries other than Japan (middle-income Korea and low-income China). In his study, Ofori (2003) then broadened the net to cover all top 225 ENR contractors in 1990–1999, and focused on Asian firms as an example:

- 4 Singapore contractors were in the list – the highest ranking was 93 in 1993
- 1 Malaysian firm was listed for three years – attaining its highest ranking of 180th in 1994
- 4 Taiwanese contractors were in the list – with the highest ranking of 97 in 1990
- 1 Hong Kong contractor was on the list continuously after 1995 – its ranking rose steadily from 183 to 63 in 1999

- Some 19 Korean contractors were listed at least once in the ENR top 225 during 1990–1999 – the highest ranking was 12th by Hyundai Engineering & Construction Co. in 1998
- As many as 35 Chinese contractors were listed at least once in the ENR top 225 during 1990–1999 – the highest ranking was 24, by China State Construction and Engineering Corporation in 1999
- 2 Philippine contractors were listed at least once in the ENR top 225 firms – with the highest ranking of 139th in 1993, and finally
- 3 Indian contractors were listed at least once in the ENR top 225 firms – with the highest ranking at 147th in 1995.

Several authors highlight the potential contribution that foreign firms can make to the development of the construction industries in developing countries through the transfer of technologies (see, for example: Abbott, 1985; Moavenzadeh & Hagopian, 1984). However, efforts to achieve this – including mandatory joint ventures between local and foreign contractors – and the requirement to subcontract minimum proportions of projects to local firms have not been very successful (see, for example: Carrillo, 1994; Ofori, 1996).

Thus, globalisation offers advantages and opportunities for construction industries in developing countries, but also poses significant problems and challenges. The key issue is for the industries to take advantage of the general trend of globalisation and its various consequences to develop their capacities, capabilities and prospects for the future. There is considerable scope for this, as there are prospects for mutually beneficial strategic cooperation among the local and foreign firms. However, there is a need for new thinking on the parts of firms from both the developing and industrialised countries, and novel enabling frameworks from governments.

Private-sector participation

Some features of, and developments in, construction which are considered in the literature as positive factors, and welcomed as such in the industrialised countries, may not be so beneficial in the developing countries. Data published by the World Bank (2004b) show an increasing trend of private-sector participation in infrastructure projects in developing countries. In various reports, the World Bank has urged these countries to use such funding mechanisms to meet their backlog of infrastructural needs. However, these countries lack the framework for ensuring that such projects will be successful. For example, the governments of the developing countries are not in a position to conduct a proper evaluation of the privatised projects. The assessment of their own risks from such projects is also beyond these countries. Thus, these countries may not obtain the best deal, and may not adequately secure their interests. Ofori (2004b) discusses these issues, and suggests that efforts should be made to align the interests of construction project participants to educate developing country authorities on privatised projects and to undertake research on the risks faced by developing country clients and consumers on privatised infrastructure projects. Another issue is that, as shown in Table 17.2, such private participation tends to exclude firms from the developing countries as they lack the track record and expertise to compete effectively for such works (Raftery et al., 1998).

Private-sector participation in public infrastructure and other projects has other implications. The government's role as client is identified as a key feature of the construction industry (see, for example: Hillebrandt, 2000). Many authors on the subject of industry development, such as Turin (1973), urged the governments of developing countries to use their bargaining power resulting from their large share of total investments to effect desirable changes in the construction industries. With increasing private-sector participation in projects, the role of governments is reducing. As governments are no longer main beneficiaries of an improved construction industry, the industry cannot expect the same level of interest and support from a government in relation to appropriate policies, programmes and actions on its development.

Cultural fit and effects on performance

With the liberalisation of the economies in most developing countries, the clients of the construction industries in these countries, which are increasingly private, are demanding better performance from the industry for less money or seeking what they consider to be 'value for money'. With the rising importance of good governance and accountability, the public client is also dissatisfied in many countries, and is demanding better results.

In the developing countries, industry structures, systems, practices and procedures remain those that were introduced by the former colonial countries, although these developing countries have been politically independent for several decades. These systems and practices were developed in accordance with, or in response to, cultures, value systems and market imperatives that are different from those of the developing countries. Even more serious, efforts have been made to change these in the industrialised countries. For example, many developing countries which are former British colonies use standard forms of contract which are similar to, or the same as, the 1973 edition of the Joint Contracts Tribunal form. In particular, many public-sector forms of contract have remained the same as those drafted by the colonial architects and engineers.

The project procurement and administrative arrangements in use in developing countries have also been inherited from Western countries, which have a different history, culture, collective experience and breadth of construction expertise. These arrangements determine the documentation, procedures and practices in the industry, as well as the roles of the participants and the relationships among them. In general, the present arrangements stress formality and rigid channels of communication. Ironically, in the countries of origin of these procurement arrangements, these arrangements have been changed, or are under discussion. For example, in the UK construction industry, Latham (1994) advocated the building of trust and a spirit of partnering in an industry characterised by mistrust, rivalries and adversarialism. The 'traditional' procurement approach where the architect is the principal consultant who leads the design and construction teams, which is still predominant in the developing Commonwealth countries, is now only one of many possible approaches in the UK.

Furthermore, the influences on the UK construction industry have come from other industrialised countries, mainly the USA. In the UK, there is now an active search for models that favour strategic long-term relationships among the participants. The more adversarial and legalistic procedures, such as arbitration and litigation, are being replaced with adjudication and various modes of alternative dispute resolution. Thus, there is scope for the developing countries to attain significant progress by considering the new approaches in these industrialised countries for possible adaptation and application.

In construction, the relatively new studies relating to culture are of direct relevance to developing countries. The studies have, so far, concerned:

- The impact of the nation's culture on construction activity (Rwelamila et al., 2000)
- The culture of the construction project (Rowlinson & Root, 1997)
- The culture of the construction firm (Liu & Fellows, 1999)
- The culture of the construction site (Applebaum, 1981)

A study by Rwelamila et al. (2000) on some southern African countries suggests that the failure to consider the culture of the nations in procurement arrangements might be one of the contributory factors to the poor performance on construction projects in developing countries. Ang & Ofori (2003) suggest that the cultures of 'the East', and China in particular, are conducive to the successful application of partnering in construction.

Given the uniqueness of culture to particular groups of people, and its pervasive influence in societies and organisations, these studies confirm that the construction industry and its practices and procedures must differ in every country. Thus, it is necessary for effort to be made to devise practices, procedures and relationships that are suited to the culture of each country. In particular, effort should be made to formulate procurement approaches that enable and facilitate the integration of the construction process in the context of the country concerned.

Discussion

The comprehensive studies of national construction industries that have been undertaken in some countries, such as Australia, Hong Kong, Singapore, Sweden and the UK, have led to the formulation of medium-term strategies for developing these industries and improving their performance. Some significant achievements have been realised in some of these countries. It is necessary for similar studies to be undertaken in developing countries. These studies should involve all the stakeholders of the construction industry, but governments should provide the lead. Greater attention should be paid to the implementation of the strategies and recommendations outlined. During the study, the championing and monitoring of the implementation process, and periodic continual review of the industry development programme, should be given due attention during the studies.

The construction industries in developing countries must endeavour to attain their potential in the key role they can play in the economies and in national development.

They must be active participants in the national effort to attain economic prosperity. In every country, the industry must make the effort to be seen as a creator of wealth for the nation.

While the trend towards deregulation and policy liberalisation is generally welcome, many areas and aspects of construction will have to remain under government control owing to their implications for public health and safety. Thus, there is the need for appropriate policies and administrative frameworks in each country. A government's role as an enabler should be given greatest consideration, and further developed. This requires governments to have greater knowledge of the nature, problems and needs of their construction industries. The education of the government should be an objective of the industries.

Procurement

In the industrialised countries, there is now stress on the choice of appropriate procurement strategies, with focus on integrated approaches such as design and build, and long-term relationships such as partnering. There is also a move away from the lowest price-based selection of successful tenderers. The developing countries should continue to monitor the research and experimentation on procurement approaches. However, they should seek appropriate systems that suit their culture and business traditions. It is appropriate for research to be undertaken on a panel of criteria for contractor selection other than price, which is suitable for the context of each developing country. Partnering would appear to have scope in these countries, as business relationships in the construction industries are, as yet, less adversarial than in the industrialised countries.

It is necessary to reconsider the arrangements and procedures of the construction industries of developing countries. The construction practices and procedures of these countries must reflect the cultural attributes and values of individual developing countries. Studies on the culture of construction and construction-related firms, projects and workers in the developing countries would help project managers to integrate the contributions of the project participants most effectively. It would enable clients to appreciate how to offer incentives, and steer construction firms, to deliver the best possible product and to innovate. It would help managers in both contracting and consultancy organisations to understand how to communicate with, and motivate, their workers. Finally, the results of such studies would also provide insights into the most effective way to transfer technology to local construction firms.

Research into procurement processes is required in order to develop innovative approaches which can engender unity of purpose and collaboration among the participants, and the integration of their contributions.

Project management

The construction industries in most countries can also observe better practices adopted by their counterparts elsewhere, and are adapting these to improve their performance and prospects. The industries in developing countries should continually monitor trends in the industrialised countries, and draw lessons from them. It is important for the industries to develop mechanisms to enable them to do this most effectively. These would include gatekeeping to identify the new developments, adapting the new concepts or procedures to suit the circumstances of the subject countries and monitoring the results of the application of these measures.

The increasingly faster pace of change in every sphere of life (in economic matters, in technology, tastes, political issues, and so on) is posing problems to construction companies and practitioners. It is also offering new opportunities. However, it requires new corporate approaches, networks and modes of operation.

There should be greater emphasis on the client's needs and the user's requirements and perspectives. In the developing countries, where the majority of the populations earn their living from the land, the impact of development projects should be assessed by taking into account the views and concerns of as wide a constituency of stakeholders as possible. The key players in the construction industry – in particular, the design consultants and contractors – should identify and focus on meeting the objectives and aspirations of the wider stakeholders of the construction process. This will make the industry continuously relevant and give it a caring reputation, thus enhancing its image. So far, several major construction projects, especially infrastructure items such as dams, have been the subject of much controversy and public protest.

The participants in the construction process should adopt continuous improvement as both individual and corporate policies, and develop practical ways and means to achieve this in their normal activities as individuals and on projects. In these regards, techniques such as value management and whole life-cycle assessment are useful. These methods need to be applied in a pragmatic manner, with less stress on format and structure, and greater concern to substance and practicality in the context of each project.

Efforts to align the interests of construction project stakeholders in developing countries should be part of the project planning and implementation process, for example as the subject of a workshop involving the key participants in the project.

It is critical for capacity in the management of project implementation, especially privatised works, to be built up among public administrators in developing countries. Key in this respect is an understanding of the risks that their countries face on projects, and the negotiation skills they require to represent the interests of their governments and citizens. It is also important for managerial and technical expertise to be built up to enable local firms to operate the privatised projects upon their transfer to the government at the end of the concession period. Such capacity building programmes could be funded by the World Bank and the regional development banks.

Industry development

The construction industries in the developing countries should adopt a long-term perspective at both the corporate and industry levels. They should explore the merits of cooperation and alliance formation (at home and abroad). At the industry level, they should seek to form umbrella bodies to play a championing role in the development effort. Such umbrella organisations would embrace the professional institutions and trade associations of the relevant stakeholder groups in the industry, and should speak with one voice in representing the industry's interests, needs and concerns to their government. A joint public–private championing effort should eventually be forged.

Each country has its own national characteristics and strengths. In each country, efforts should be made to identify and 'develop' these strengths, and utilise them in the effort to develop the construction industry and improve its performance.

Construction enterprises from developing countries should be enabled to participate in the international projects undertaken in their home markets. The most appropriate approach to the development of this capability will differ from one country to another as national contexts and cultures are not the same. Previous attempts to enhance capacity in local construction contracting enterprises, such as mandatory joint ventures and sub-contracting portions of projects, have not succeeded and more innovative schemes are required. Getting the potential beneficiaries involved in the design of the programmes would help.

Education

Education can play an important role, which was also the conclusion of the Construction 21 study in Singapore (Construction 21 Steering Committee, 1999). The developing countries should study models elsewhere. There is much change in many countries: for example, the UK introduced four-year engineering degree programmes a few years ago. Continuous learning is important if the construction industries of the developing countries are to keep pace with developments in more economically advanced countries and improve their performance. In view of the limited resources available to them, the developing countries should ensure that formal education is of the right type. Moreover, continuing professional development should also be a norm. Firms must also adopt a more strategic approach to the management of their human resources, and endeavour to keep, continuously develop and rationally deploy their people.

The client representatives, especially on privatised projects, require education in order to effectively engage appropriate consultants, manage their individual contributions and interpret their findings and recommendations.

In the developing countries, there are strong links between the universities and the construction industries. This is because several senior academic staff members are also prominent in the professional institutions, several of whom have thriving practices. These relationships should be further developed and utilised towards the development of appropriate educational programmes, and the enhancement of the performance of the construction industries.

Efforts should also be made to form relevant regional (international) umbrella groups of the construction industries, which will enable the countries to share experiences and ideas and to launch and implement mutually beneficial joint efforts. There are regional and global groupings of individual professions and trades such as the African Union of Architects and the International Federation of Asia and West Pacific Contractors Associations. While greater effort could be made to utilise these existing groupings to maximum mutual benefit, regional groupings with wider representation, preferably at whole-industry level, would be useful. These could be patterned on regional political or economic blocs such as the Association of South-East Asian Nations (ASEAN) or the African Union (AU).

Conclusions

For developing countries, it is important for the construction industries to be improved owing to their critical role in the national economies, and to long-term socioeconomic development. These industries have been taken for granted by governments and societies, and have failed to attain their full potential. Indeed, their performance has been unsatisfactory from the point of view of clients and users, as well as governments. It is time to revalue the industries in order to improve their performance and enhance their image. This revaluing exercise should involve all the stakeholders of the construction industries.

The industry should view its future with unity of purpose and appreciation of the mutual benefits accruing from the effort to revalue the industries. There is hope that this orientation can be achieved because the rivalries and antagonisms among the participants of different professional backgrounds are not yet as entrenched as in the industrialised countries. The industry should have a long-term, strategic orientation. The proposals and recommendations formulated as part of the revaluing exercise should be country-specific, and take account of the cultural and resource contexts, as well as the governmental mechanisms and the business networks. To realise the successful implementation of these recommendations, a championing effort is necessary.

Part 4
Implications in Practice and Conclusions

18 Exemplars of 'revalued' construction

Lucinda Barrett

Introduction

This chapter provides a glimpse of how construction could, and should, be based on exemplary project examples. The work reported upon here draws from 17 mini-cases and 3 fuller cases elicited from the workshops carried out in five countries. A total of 40 senior individuals participated in the workshops, representing a good cross-section of the interests around construction: 35% had a background as clients; 38% as consultants; and a further 27% came from contracting. These 40 individuals each spent a full day working in detail as described and as such are a powerful international voice.

At the start of the workshops participants were told that at the end of the day they would be asked to give an account of the construction project or activity that for them best exemplified the creation of value. So, at the end of the final sessions, infused by all of the preceding phases, attendees were asked to volunteer their examples and the reasons for their choices. A total of 17 individual examples were given, summaries of which are given in Table 18.1 together with an indication of their connections to four emergent themes, which are themselves summarised in Table 18.2. Later on in this chapter more detailed case examples have been provided by collaborators, followed by a discussion section and conclusions.

Synthesis

Working through the examples in Table 18.1, Examples 1 and 2 show that enhanced collaboration or innovation alone can have significant impacts, reflected in phrases such as: 'all involved gained' and the 'system worked incredibly well'.

The incidence of these isolated themes is, however, the exception and it was much more common for the examples to display a combination of collaboration, or collaboration and creativity, *driven* by severe constraints (Examples 3–6 and 7–9, respectively). This seems to reflect situations where those involved are under great pressure, but through collaboration, sometimes with creativity added in, the challenge is successfully met. Sentiments are expressed, such as: 'the whole team simply pulled together'; 'the feeling of working in such a way was great'; 'initiative and energy of colourful and

Table 18.1   Examples of exemplary projects from workshop attendees
Table order from simple to complex.

Workshop attendees in the US, Australia and Singapore were each asked to indicate the project that for them exemplified the creation and delivery of high value by construction, with reasons. The following examples resulted:	Source	Emergent Themes			
		Constraints	Collaboration	Creativity	Community
1 D + B office development carried out by client and construction team working closely together. Construction was excellent and cost very low, with all involved gaining.	Sing		▨	▨	
2 A particular example was given of a project in 1972 that exploited very successfully pre-cast concrete technology such that a 7500 sq ft building a week was built. The modular system worked incredibly well, but why not everywhere now?!	US			▨	
3 A high-profile public project with very many constraints of cost, time, quality and security unusually went forward with the *second* lowest bid. Thus, the project supported a reasonable profit margin for those involved and this appears to have been significant in it running extremely smoothly and completing ahead of time.	Sing	▨			
4 A US $700m housing project in Saudi was completed three days early, on only 24-month programme and under budget by adopting a turnkey alliance approach.	Aus	▨	▨		
5 An off-shore 'hard dollar' project faced a two-day window within which a connection had to be made, and as a result the whole team simply pulled together. The feeling of working in such a way was great.	Aus	▨	▨		
6 Project in major shopping area with very severe cost, disruption and time constraints was negotiated with only two contractors, resulting in a guaranteed minimum price but no exposure on variations. Tenants and users of shopping centre were almost unaware of the works and no mistakes were made.	Sing	▨	▨		
7 Another personal example was of a school project being completed against the odds in the highly restricted Summer break through the initiative and energy of colourful and talented builders.	US	▨	▨	▨	
8 A very fast project in Darwin for 40–50 steel frame buildings to be built in 18 weeks including land acquisition demanded that (very unusually) all four competitor suppliers in the area worked together. Because of the novel approach taken and the overriding pressure a very successful outcome resulted.	Aus	▨	▨	▨	
9 A Queensland museum built in 1997 was achieved very quickly and successfully by discarding the traditional approach, greatly simplifying the processes and creating a small project team, including the contractor.	Aus	▨	▨	▨	
10 The project director on a condominium project 20 years ago was very impressive in the way he creatively addressed the project issues and found innovative solutions. For example, the estimate for the project was $190m, but in fact by calling the consultants together and working through the practicalities this was reduced to $156m. Under his leadership, this active approach to finding optimum solutions continued throughout the project in a multitude of small ways with big, cumulative impacts.	Sing		▨	▨	

Table 18.1 (*Continued*)

Workshop attendees in the US, Australia and Singapore were each asked to indicate the project that for them exemplified the creation and delivery of high value by construction, with reasons. The following examples resulted:	Source	Constraints	Collaboration	Creativity	Community
11	The 3.8 km Melbourne City Link tunnel leaked and to fix it a moratorium on disputes was put in place and the energy and imagination released in finding a solution was incredible.	Aus		▓	▓	
12	An alliance approach to the Georgina River Bridge brought the Aborigines within the alliance and transformed their orientation as the project impacted on a sacred site. The Aborigines provided labour and worked very hard to achieve the project creating ownership in the bridge, which became part of their 'dreaming'. Although a practical project it also achieved elements of social reconciliation.	Aus		▓	▓	▓
13	The creation of a major laboratory by a small focused team that managed to successfully represent both immediate and on-going perspectives.	US		▓		▓
14	The US railroad/telegraph system was proposed as an example of private sector endeavour that changed the face of society.	US			▓	▓
15	The political skill in Congress of both getting buy-in to the US highway system affecting the whole country, and at the other extreme the large investment in the 'Big Dig' just in Boston.	US				▓
16	Against a lot of pressure, time was taken out to fully work through the issues around creating age-care accommodation, including generic designs and prototypes. Although this took a lot of time, the programme of building is now proceeding very well and the investment of time has been amply vindicated.	Aus	▓		▓	▓
17	Several participants cited the 'Big Dig' in Boston because: It will deliver value for the next 100 years; It has been technically and aesthetically successful; It meant dealing will multiple stakeholders, but was handled very transparently and openly; Hundreds of historic buildings were involved/protected; Claims were radically reduced through innovative, real-time monitoring; Highly collaborative, non-combative ways of working.	US	▓	▓	▓	▓

Table 18.2 Emergent themes infusing exemplary project examples.

Theme	Brief description
Constraints	Generally resources and often time in particular, but sometimes issues such as physical limitations on the site. Often 'supernormal'.
Collaboration	Unusually close working together, typically involving the client, designers and contractors from early on in the project.
Creativity	Can be an innovative technical solution or procurement approach, but normally reflects a strong joint problem-solving approach throughout.
Community	Where the project positively impacts more broadly than the immediate participants in its long-term orientation and/or broader societal engagement.

talented builders'; 'very unusually' and 'discarding the traditional approach'. These convey the feeling of strong social bonds flourishing once the stifling limitations of 'normal business' are relaxed in the face of extreme demands.

It is interesting to note that in several cases of constraints, a way forward is fashioned by actually releasing a significant limitation. For example, taking the second lowest bid on a project that was very high profile (Example 3), guaranteeing a minimum price (Example 6) and explicitly taking time out to fully work through the briefing/design issues (Example 16).

Examples 10 and 11 highlight the potential of close collaborative working to actively stimulate and *support* ongoing creativity. The impact on those involved is again indicated in the comment that: 'the energy and imagination released in finding a solution was incredible'.

Collaboration and creativity can both be seen to *support* wider 'community' impacts (Examples 12–13 and 14–15, respectively). The case of Example 12 is a rich example in which instead of tension with indigenous people, they became a key part of the project and it became 'part of their dreaming'.

The final group of examples (16–17) exemplify projects in which all the themes are evident. As Case study 2 reveals, maybe it is the huge scale of the 'Big Dig' in Boston that leads to this rich scope, but the theme of *ambition* is evident in other examples too, in terms of strong environmental aspirations and treating the project as a showcase.

Based on this material, at its richest, 'revalued construction' can be typified by projects where significant constraints drive those involved to collaborate strongly, spurring the team to innovative responses which not only triumph against the demands of the project itself, but also impact positively on the community around. Thus, it is suggested that the emergent themes link together as shown in Figure 18.1.

To test the '4Cs' model, three case studies were developed in more detail through collaboration with three individuals who had been directly involved in the specific projects.[1] This led to more detailed illustrations of aspects of the model in practical terms and, through the discussions to a clearer understanding of the four categories themselves. The three case studies are set out below.

Case study 1: William McCormack Place, Cairns, Australia

Brief project overview and measures of success

William McCormack Place is a four-storey commercial building of 4568 square metres developed by the Queensland Department of Public Works in order to provide

---

[1] Thanks are due to those who provided case study material: Graham Messenger for Case study 1, Thom Neff for Case study 2 and Jacqueline Vischer for Case study 3.

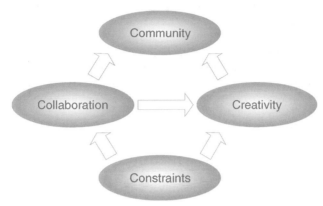

Figure 18.1   Dynamic '4Cs' model.

Figure 18.2   William McCormack Place.

sustainable office accommodation. A private sector construction manager, with an overall budget of A$17.5m, constructed the building. The final project exceeded the environmental impact targets that were set. Further, it was completed on time, at A$400 000 below the business plan budget.

Constraints

Strict environmental sustainability targets under the Australian Building Greenhouse Rating Scheme demanded that the project had to meet at least a four-star level. The rating that was achieved actually exceeded the five-star level by 15%, equivalent to 317 tonnes of greenhouse gases saved. Such energy performance savings for the owner worked out at A\$6.75 sq m/annum as opposed to the projected A\$5 sq m/annum.

The project had also to minimise the environmental impact throughout its entire life-cycle, whilst still being a commercially viable unit that could flexibly accommodate a high churn rate of users. In terms of the actual construction work, a requirement was made that particular attention should be paid to adverse environmental issues, pollution in terms of dust, mud, noise and water run-off.

Collaboration

In order to incorporate and deliver the environmental sustainability targets it was seen as essential to have honest, open business relationships and shared vision between the various stakeholders. The make-up of the project team reflected this objective. Overarching control of the project was in the hands of the facility management team that was to have responsibility for the management of the building in the long term. Further, staff of the Environmental Protection Agency, as a major tenant, were also engaged in the design work. Local indigenous cultures were also involved in the project, with large-scale Aboriginal works promoting community ownership of the building.

Creativity

The constraints set by the environmental targets stimulated creative responses in the design team. For example, this can be seen in relation to the design and operation of the air conditioning unit. The inclusion of a thermal wheel enabled a 25% reduction in the size of the refrigeration plant and consequent energy savings. These innovative designs will also result in large life-cycle costing savings in maintenance and capital replacement. The design of the external structure incorporated long-lasting, low-maintenance materials. Internally, the fit-out was also driven by the need to be ecologically sustainable in order to maximise the flexibility of the building throughout its lifespan. For example, workstations were designed to use recycled components and mechanical fixing to facilitate dismantling. The design of the building as a whole will enable the majority of its façade and internal components to be recycled. Resistance to budgetary pressures and an unwillingness to compromise on the environmental goals stimulated these project design solutions.

Community

In line with the sustainability agenda, the importance of economic sustainability in the local community was not overlooked. The project used local firms and materials and, in addition, many will be used in the day-to-day operation and maintenance of the property. The building also contributes to the social environment of the area, with landscape, planting and art works used creatively to make the building feel accessible, inclusive and inviting for the community.

Figure 18.3   'The Big Dig'.

Case study 2: Central Artery/Tunnel (Ca/T) Project, Boston, USA

Brief project overview and measures of success

This 20-year, mega-project was fully completed in December 2005. When it began in late 1985, it had already seen more than 15 years of political and economic controversy and resistance, and was considered by many to be 'impossible' to carry out success-fully. It started under the direction of the Massachusetts Highway Department and eventually was taken over by the Massachusetts Turnpike Authority. The owner chose Bechtel/Parsons Brinckerhoff (a joint venture) to carry out the tasks of: permitting; en-vironmental impact assessment; preliminary design; management of final design (by others); and management of construction (conventional design–bid–build by others). It is essentially impossible to fairly assess the 'success' of the project based upon con-ventional scope and schedule criteria, as the final works do not resemble the plans envisioned in 1985. The project revitalised the entire city of Boston, created a new 3-mile linear urban park, kept traffic moving throughout the construction process and raised the value of taxable real estate along, and adjacent to, its alignment by billions of dollars. It created a totally new and exciting waterfront along Boston harbour. In the words of our collaborator on this case study: 'Only an irrational critic would say that the CA/T Project was not a success', but with any scheme this innovative there will always be problems to be resolved.

Constraints

Most of the project had to be built directly beneath an ageing, elevated highway that, in turn, had to be underpinned and kept fully operational during new construction.

The project traversed a congested part of the city, through largely historic fill materials with countless obstructions, environmental contamination, notorious poor soil conditions and high groundwater levels (that had to be left unchanged). In addition to groundwater protection – limiting of deformations related to construction, controlling environmental contamination issues, and the maintenance of traffic and people access – things like dust, vibrations, noise, air pollution and aesthetics were placed into the mitigation mix. The alignment passed adjacent to, or directly over, four active subway lines, two active rail terminals and countless buried major utility lines (some of which were more than 100 years old). The alignment also passed by hundreds of existing buildings and other facilities, some of which were historic in nature, founded on wooden piles, or as close as 12 feet from a planned 100-foot deep excavation. Permitting and mitigation involved numerous local, state and federal agencies, as well as numerous citizen action groups and private facility owners (and their consultants). Excavations of immense depth were required, crossing Boston Harbour and the Ft. Point Channel (with immersed tube tunnels), crossing the Charles River with a bridge and interfacing with Logan International Airport, as well as sensitive manufacturing facilities that permitted essentially 'zero' settlement. The entire design and construction process was driven by four key issues: protecting the existing groundwater levels; limiting all deformations to very small magnitudes; avoiding any adverse environmental issues; and keeping all traffic and pedestrians moving freely.

Collaboration

The noted constraints demanded an unprecedented level of collaboration in order to advance and complete the project. The massive environmental impact statement (EIS) set the tone for the project. The owners went to great lengths to address all potential adverse impacts and put in place a process to both address and resolve them at the earliest possible time. Formal political, community and business liaison groups were put in place and supported throughout the design and construction process. These groups met on a regular basis for two-way dialogues to explain the intent of the EIS and the design, and to react quickly to issues raised by involved entities. Collaboration was encouraged between and among the large number of involved agencies: local, state and federal environmental offices; the Corps of Engineers; the Federal Highway Administration (FHWA); the Massachusetts Historical Society; and others. For such a lengthy project, important 'lessons learned' from early contracts were incorporated into later design and construction efforts to a great advantage. Formal 'partnering' efforts were utilised throughout for both design and construction firms to help ensure buy-in to the project's overall goals and objectives.

Creativity

The numerous and serious project constraints, coupled with an almost unprecedented level of cooperation (driven by formal 'partnering' efforts) led to a high level of creativity during the design and construction processes. Miles of slurry walls were constructed, some as deep as 120 feet, to both provide cut-offs to control groundwater and to provide very rigid, lateral structural support to control deformations (most sections were 36-inches thick and utilised 36WF300 structural members as reinforcing). Carefully staged excavation and extra lateral bracing were also employed to ensure minimal deformations resulting from construction.

A number of techniques were employed throughout the project to permit safe construction, or to overcome seemingly impossible field conditions. These techniques included: ground freezing; large jacked box tunnels; immersed tube tunnel construction (concrete and steel); jet grouting; deep soil mixing (largest volume in the USA); groundwater recharge (first successful use in Boston); a non-symmetric cable-stayed bridge; jet fan tunnel ventilation; and an extensive monitoring system (some in real time) that included all sensitive and historic facilities adjacent to the alignment (utilities, subways, structures, groundwater levels, noise, vibration, etc.). In several cases, creative collaboration between agencies permitted new subway construction to interface with ongoing CA/T construction to the benefit of both entities. The monitoring system, that at its peak had 5000 data points, was so successful at providing timely data on critical variables, that in some cases construction was 'controlled' to prevent potential damage, or lost time and construction claims.

Community

Throughout the entire process, from environmental assessment through construction, formal community liaison teams were employed to both solicit input and to react quickly to the development of potentially adverse situations. By choice, the final design and construction contracts were broken into packages of a size to permit the extensive involvement of local firms. The project included the removal of an ageing, unsightly elevated highway that was a barrier between the majority of Boston and its waterfront, and the creation of a linear urban park as well as a new exciting 'gateway' to a revitalised waterfront. Landscaping, planting and artwork will be incorporated into the park (again, with much community input). The value of the taxable real estate along the alignment has increased by a few billion dollars, creating new revenue for the city to the benefit of many citizen initiatives. The buried highway will permit through-traffic to traverse the city in minutes, rather than be stuck in the previous massive traffic jams: wasting time, wasting fuel, and creating massive air quality problems.

Case study 3: Hypertherm Inc., Hanover, New Hampshire, USA

Brief project overview and measures of success

In 1993, this medium-sized manufacturing facility, employing 350 people and producing hi-tech, metal-cutting equipment sold all over the world, expanded its manufacturing floor from 8000 sq ft to 15 000 sq ft and added a 2-storey office structure comprising approximately 30 000 sq ft. The project was initiated in a conventional fashion, by hiring an architectural firm to provide a programme (brief) and apply for zoning approval for

Figure 18.4   Hypertherm Inc.

expansion in its rural and natural setting (protecting wetlands, minimising tree cutting, protecting natural fauna, etc.). The inadequacy of the conventional briefing process for a complex change management problem led to innovative ways of making design and construction decisions, and resulted in a building considered to be a major asset to the company. A full account of this case is given in Vischer (2005, Chapter 2).

Constraints

Project constraints consisted of the limitations on form and extent of expansion imposed by the protected rural setting, a conventional budget and the conventional approach taken by design and construction professionals to creating the new work environment for Hypertherm. The CEO rejected the brief submitted by the architects on the grounds that the rapid growth of his company required more than simple linear expansion of existing square footage; a larger organisation required a different structure to be successful. The construction company offered design services, but discouraged innovation on the grounds that any unconventional aspect of the building would cause costs to rise.

Collaboration

The architectural firm responsible for providing the initial brief was replaced by a small firm of work environment specialists whose mandate was both to help the company determine its new form and structure, and also to facilitate the process of making key design decisions for their new work environment. The facilitators established a process whereby senior managers agreed on their definition of the new organisational structure and they helped them communicate this to employees. A secondary process, to involve employees directly in decision-making regarding the design of their new work environment, was implemented to ensure buy-in from all levels of the company. The outcome of this extended briefing process was then communicated to the contractor, to ensure that detailed design and construction decisions continued to respect the client's new goals and values.

Creativity

The abandonment of the conventional design and construction process offered a series of opportunities for creative problem-solving. Briefing, for example, had to be redefined, and a new process was implemented to involve users at all levels. At the same time, principal decision-makers (the CEO and the senior management team) engaged in defining the new organisational structure and in setting overarching goals and design principles that subsequent decisions had to respect. In this process, the brief became a vital tool for continuing to inform and guide design as the project evolved. Design development became an opportunity to involve more end-users in the process, and gave them a say in the workspace they would eventually occupy. The process inspired a new and more creative relationship between the contractor and both the design team and the client. Through being involved in planning and managing change, the contractor participated in finding innovative solutions to technical problems that arose during construction, such as, for example, increasing ceiling height (distance between floors) in order to permit suspended indirect lighting in the office areas.

Community

Hypertherm moved into the new building in 1996. As a result of their close involvement in all stages of the process, the firm's employees took ownership of the new space and adapted smoothly to the new organisational structure and way of working. There was a feeling of being pioneers and successfully leading change, which led to a positive *esprit de corps*, in contrast with more usual employee reactions of fear of and resistance to change. The senior management team had opted to occupy the same open-plan workstations as employees, who then saw this as a commitment to the new work environment and to the egalitarian values it represented. In addition, Hypertherm employees proudly invited clients, vendors and new recruits to visit the building, considering that the new work environment was a major selling point for the company. This was positively viewed by the local community, for whom Hypertherm was a major employer, and also helped in recruiting new staff to this rather remote and rural part of the state. In the months after project completion, Hypertherm was voted the best place to work in New Hampshire.

Summary of Case studies

Looking across the three case studies it is apparent that, through the use of the '4Cs' model, it did prove possible to credibly describe the key aspects of these projects that led to their success. However, it also became apparent that it is important to explicitly cover contextual issues and, especially, the measures of success relevant to each project. With this rider, this part of the research has provided greater confidence that the model is robust across diverse sectors (in these cases: an office, a highway and a factory) on hugely different scales. It has shown that the model has a positive role to play in helping gain a relatively simple understanding of, often, hugely complex projects.

Discussion

The '4Cs' model, grounded as it is in the very best experiences practitioners have had, will now be related to the more general revaluing construction discussion that revolves around topics, such as procurement, briefing, design and construction management, together with wider aspects such as user needs and sustainability.

A core issue in the area of procurement, especially in the public sector, is the pre-eminence of the lowest cost approach. In Chapter 11 this is typified as: 'competition is good; more competition is better'. The result seems to be that this one-dimensional approach leads inevitably to a situation of 'lock-in' where it is not in the interests of any individual party to move out of their role, which itself is quite logically moulded to maximise their self-interest. This resonates strongly with the discussion in Chapter 5 in terms of the 'prisoner's dilemma' analogy, in which through rational choice each aims to gain, but as a consequence all lose. The solution to this dilemma centres on cooperative responses that look beyond the immediate factors in play and allow trust to develop. Partnering/alliancing arrangements are a clear move in this direction.

A dynamic balance is the key, and here the choice of major constraints or challenges for the whole construction coalition is highlighted as a pivotal aspect by the '4Cs' model. This involves the careful identification by clients of what are the significant things that matter to them, driven by their particular circumstances, and the communication of this to other participants in the project, whilst being very flexible on other dimensions so that those involved have the scope to respond. The examples given exemplify how this typically creates a very demanding situation, whilst not simply relying on 'least cost' as a blunt instrument to control the parties involved. This calls for flexibility and finesse to ensure that the appropriate approach to procurement is taken to achieve optimal results in the particular circumstances that surround each project.

Within the context provided by this sort of procurement approach, unusual modes of collaboration emerge. Typically, these involve close interaction between many of the construction players from an early stage, but crucially also place the owners/users centre stage, often alongside outside societal stakeholders. This broadening of those involved towards a life-cycle perspective, rather than a snapshot project view, links well with the partnering ideas mentioned above, but also with the notion of value management (VM) introduced in Chapter 6. There is a connection too with views in Chapter 8 on assessing building performance, which articulates with the notion of *fit* between what the client wants and how it is provided by the construction industry. Achieving that fit under the extreme pressures of project conditions is difficult and again the clear identification of client priorities can help enormously, but demands a close interaction between those parties that extends through the project beyond just the early stages.

Interestingly, the three more detailed '4Cs' examples given show that close, inter-active working throughout holds many benefits, not just for the fit with a client's requirements, but also at the level of technical, project-based innovations in support of those broader aspirations. The literature on briefing touches on these issues too, with Barrett & Stanley (1999) stressing the importance of conceiving this process as running dynamically through projects and that the clients should be 'empowered' to clearly express their needs related to their operations. Horgen et al. (1999) reinforce this by stressing the iterative, non-linear, social/political nature of briefing in practice.

This interactive, broad grouping of people working around and towards the project – or better, the client's – objectives has implications for the design process involved. Zeisel's (1981) work makes a strong link with the notions of evolving, iterative design linked to user needs, expressed in Chapter 6. Lawson (1990) stresses that architectural design is often performed in groups, with designers 'hammering out rather than easily conceiving their ideas' (p. 179). Moreover, Duffy (with Hutton, 1998) has long argued, from within the profession, that architects should actively seek such collaboration. In construction there is clearly a significant overlap between various professionals, such as architects, structural engineers and services engineers, as well as great potential along the axis running from clients to contractors (Gray & Hughes, 2001; Spence et al., 2001). Along with these authors, Lawson (1990) highlights the difficulty of holding this sort of temporary team together, but also states that when it gels it can generate 'the greatest satisfaction man can achieve: creative and productive group work' (p. 194). This resonates with the '4Cs' cases where it was clear that those involved had experienced quite profound pleasure from the projects, despite the pressures, and seemingly because of the feeling of involvement in a severe joint effort to address a very testing challenge. For this to happen, creating the right conditions is important, which relates back to the procurement and flexible working issues discussed above. It also reflects, at a project level, many of the tensions around creating and sustaining shared visions covered at a national level discussed in Chapter 4.

An aspect of these conditions and the sources of pleasure, in some of the cases reviewed, came from engagement with and the satisfaction of potential users, and also from broader societal actors. Construction has widespread impacts as set out in Chapter 8. Buildings contribute to urban spaces and regeneration on multiple fronts, with profound organisational and societal consequential impacts, but construction also carries significant environmental impacts. From the '4Cs' cases it can be seen that many of these issues are touched upon in the projects cited (environmental issues, user performance, local employment, art in the community, etc.). It can be speculated that part of the pleasure that made these projects stand out was derived from the engagement of those involved with this richer range of actors, extending well beyond the normal project horizon.

Conclusions

The examples of exemplary projects cited by the workshop participants are typified by the '4Cs' model, namely constraints driving collaboration and creativity, ideally lead-ing to community benefits. The more detailed case studies illustrate that the model is a useful generic tool to describe the major aspects at play and the interactions that made the projects so memorable. The model stresses connections between the parts as experienced in the real world and so provides some indications of bridges between these issues. For example, strong connections between procurement-driven enhancements in broadened collaboration and issues in the areas of briefing, design and sustainability, not to mention the satisfaction of those involved.

A major aspect within all of this is the role of constraints and the interesting question, in the project arena, as to when is a constraint a positive factor and not a restraining force? The answer would seem to be when it is stated explicitly, clearly and early on. Further, when it is demanding enough to define the project, prioritise and re-orientate behaviour around a super-ordinate goal and provide a clear measure of success. This is aided by a situation where other (less important) parameters are dealt with flexibly so that appropriate collaboration and creativity, both technical and organisational, is facilitated to meet the challenge. In this way the constraint has provided a clear space to work within and this certainty can clearly be stimulating, provided sufficient flexibility with the remaining resources is available. This contrasts with most restraining forces that sap energy and hold back initiative and hamper creativity. Another pertinent aspect is the nature of the constraint itself. Far from being arbitrary, participants clearly understood and accepted the rationale for these restraints, whether social, time or environmental. Success with such a task is then explicit and the pride of the participants in these projects was clearly evident as they told their stories.

The approach of requesting these exemplary examples worked better than we could have hoped. The exemplars given are typified by groups of people facing demands that acted to break the dysfunctional paradigm of normal practice, so allowing refreshing and motivating actions to follow. This provides a glimpse of how, arguably, construction could and should be. However, the more detailed case studies highlight the necessity to consider each project in its specific context, and so a balance is needed between the opportunities highlighted by the generic '4Cs' model and the finesse needed to take these up in practice.

The work reported upon here covers 17 mini-cases and three fuller cases all drawn from three countries. The value of these cases lies in their richness and connection to real projects and the multiple factors at work, and as such they can provide a very valuable complement to more focused, theoretical works. However, further work to build additional cases from a wider range of countries will test the broad outline of the '4Cs' model and increase our understanding of the dynamic mechanisms at work.

19 Stakeholder action areas

Peter Barrett

Introduction and context

So far this book has set out, and expanded upon, the seven key areas that demand combined attention to progressively revalue construction. The previous chapter has indicated the complex interactive relationships that underpin projects in practice. The question remains, however, as to *who* will own the required actions.

The workshops held in five countries supply a rich context for consideration of this question. This present chapter provides an analysis that first sets out the context for action and then proposes key stakeholders for bundles of the actions such that together all aspects are covered.

The first step in the analysis is, therefore, to consider the various driving and restraining forces across all of the workshops. These are shown in the 'map' in Figure 1.3 (p. 6 of this book) and divide into three broad types, as below:

- **Frameworks** – These are factors that provide the broader structural or organisational context
  - Examples of restrainers are: lack of structure for innovation; limited land space for construction; lack of continuity and fragmentation; and unbalanced government rules and regulations
  - Examples of drivers are: government legislation; better processes; innovation in materials; population growth; email and ICTs; and long-term relationships

- **Knowledge/attitudes** – These are general factors, but those that attach to individuals
  - Examples of restrainers are: poor image of construction; inability to manage or transfer risk; uninformed team members; training/education deficiencies; society conservatism; and short-term financial client orientation
  - Examples of drivers are: knowledgeable and focused clients; credibility/keeping promises; intrinsic satisfaction; emotional intelligence; and demonstrable cost/value

- **Project** – These are factors that collect specifically around the project phase
  - Examples of restrainers are: system of lowest bidder; limited budgets and other parameters; inappropriate contracts; lack of stakeholder participation re operations and maintenance; and 'cut throat' competition
  - Examples of drivers are: single point procurement; clear communication of needs and perspectives; shared common goals across the team; early contractor involvement; and thorough programming and research in the early stages

The drivers and restrainers are shown on the 'map', and this begins to give some feel for the 'thinking space' within which initiatives for revaluing construction must be placed. It is fair to say that, although there are variations, particularly of emphasis, between the study countries, there are also the common thrusts, summarised below.

Significant problems are evident in the absence of stability and balance in *frameworks* that could take construction beyond the current position of stumbling from project to project; however, there are opportunities to build on, such as the over-arching demands of population growth and governments' focal areas of influence. Many of the constraints on progress are seen to be to do with the *knowledge/attitudes* of all those involved in construction, and this includes their education, training and basic attitudes. However, where there are enlightened individuals these constraints are seen as leverage points for progress, leading to a collective orientation towards value and away from cost. 'Actions speak louder than words', and once a project is initiated the practical influence of the frameworks and attitudes described above becomes very clear in the prominence of the criterion 'lowest first cost'. However, drivers for development are seen around the early involvement of contractors and the broader consideration of users needs, all within more mutually beneficial relationships.

Actions

The actions then suggested by the workshop groups inhabit the space between the driving and restraining forces. Categorising the actions using the notion of stakeholder provides a strong underlying rationale, as actions without 'owners' are likely to remain un-done. The actions were put into the following categories:

- Industry/client
- Government
- Procurers
- Project team
- Education and research
- Society

The actions were tabulated showing associated drivers, restrainers and linked actions. Based on an initial analysis by country, the following can be noted:

- In the area of 'industry/client' actions there is a broad range of proposals, albeit with little from the UK. The USA has a single, but significant, proposed action. In general, there are multiple drivers, many restrainers and a moderate number of linked actions. This profile suggests the realistic possibility of successful progress in this area.
- In the area of 'Government' actions there are lots of proposals. The UK, in particular, has a lot of actions, although Australia is barely represented. There are many drivers as well as restrainers and linked actions, reflecting the perceived feasibility of this as a source of movement towards improvement.

- 'Procurers' actions make up the biggest listing, with many proposals from the USA and Singapore in particular, but very little from Australia. This was a very big area for the UK too, slightly masked by being collapsed into a joint major node on the diagram. There is a reasonable foundation of drivers, quite considerable restrainers and very many linked actions. This appears to be an area that is quite generally perceived to be a leverage point for initiating progress.
- For 'project team' actions the UK and Australia appear strongly; however, Canada is absent, the USA barely figures and Singapore is only lightly represented. There are only a few conceptual drivers, but many restrainers and linked actions, again indicating the lack of a strong base from which to act. However, this is the one area in which IT arises, variously as an action, restrainer and driver.
- In the case of 'education and research' actions Canada is dominant and Singapore only lightly represented. It is clear that there are few direct drivers, many restrainers and a lot of linked actions. In other words, education and research seem to be areas that can be of systemic importance, but they do not have a strong base in the construction sector from which to exert influence.
- In the 'society' actions area only Canada and the USA proposed actions. Issues such as 'work with unions' may well be particular to the country involved. However, the general issues related to longer term and broader factors, such as social/user engagement, should surely be of general note. Their absence in other countries' discussions is probably an indication of their low perception of the likelihood of practical success for actions in this rather abstract area, rather than the variable importance of these issues in themselves.

The picture is one in which the participants in the workshops displayed different emphases:

UK	–	Government, procurers and project team actions
USA	–	Procurers and society actions
Canada	–	Education and research and society actions
Australia	–	Industry/client and project team actions
Singapore	–	Industry/client and procurers actions

However, these are indications of the major action areas that the various groups chose as being feasible in practice. Almost all contain at least one major leverage point from the first three listed. Three stress the complementary importance of project team actions. So there are consistencies and within the drivers and restrainers there is considerable overlap, indicating great similarities between the situations in the countries involved.

Moving from this assessment, it seems reasonable to suggest that overall, at a generic level, certain elements are variously available and can work positively in concert. For instance, the industry and clients can act together to improve the environment for raising the performance of construction. In addition, governments in particular can be influential in terms of policy leadership and as a major client. At the more operational level procurers hold a key role that can significantly impact on how projects are set up. Linked to this there is considerable potential within projects to make progress, but this is quite dependent on the scope provided by procurement and other 'upstream' actions.

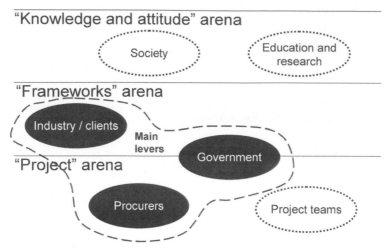

Figure 19.1 Summary disposition of stakeholders.

Education and societal perspectives are doubtless important contributors, especially taking a longer term view; however, they are unlikely to be the main, immediate focus for driving improvement.

In summary, it would appear that major levers for change are industry/client forums, governments and procurers. The other, rather dependent focal area is the project team; and, at a more abstract level, 'society' groupings and education and training have the potential to make longer term impacts. It can be argued that these stakeholders operate at various levels. Using the characterisation developed for the driving and restraining forces above, namely: the 'knowledge/attitude' arena; the 'frameworks' arena; and the 'project' arena, Figure 19.1 illustrates each stakeholder group inhabiting one of the three arenas, with the exception of 'Government' with its dual roles as policy leader and major client. Similarly, there is an overlap between clients' strategic role and their operational actions as procurers. The three main leverage areas for which there appear to be sufficient associated driving forces are shown in heavier ovals.

The next part of this chapter will look in more detail at the actual actions that the stakeholder groups can own. The discussion will highlight the major themes and connections emerging from the views from five countries. These will then be used to create an over-arching summary, re-connecting the various elements.

Industry/client

This is a key stakeholder group operating in the 'frameworks' arena. The primary action area for the industry/client grouping is to create the *capacity* for action at a strategic level within and beyond the construction industry. This was described in different ways and exists to varying degrees in the different countries. Thus, the creation of service-provider (industry) and client-focused forum/s to stimulate innovation, benchmarking and information exchange is the key action in this area. This should draw on what could be called existing 'nodes for development', namely: good long-term relationships; teams that have already achieved integrated operations; and individuals with high-level interpersonal skills. It is crucial that amongst those involved there are informed clients to ensure a strong service orientation, looking beyond internal industry issues. Such an alliance will provide a strong antidote to a confrontational industry, resulting in shared information sources, a clear voice to promote balanced Government rules and regulations leading to a conducive framework for industry innovation.

The other action areas build from this capacity on three fronts: creating integrating information frameworks; consensus 'regulatory frameworks; and joint promotional themes. In the case of information frameworks, R+D and advances in IT use are important drivers, but action is restrained by the lack of continuity and fragmentation in the industry. Consensus (but flexible) regulatory frameworks can provide an enhanced degree of stability, and when extended to include standard forms of contract they can stimulate the use of appropriate contracts and raise expectations regarding quality of documentation. Joint promotional themes must draw on examples of professionalism and achievement in the industry and should additionally exemplify the intrinsic satisfaction of much of the work of the industry. In this way the poor standing of the industry can be addressed. Throughout these actions it is important to maintain the involvement of knowledgeable clients so that the service-orientation and contribution of the industry is kept in focus.

Government

The government stakeholder is separately identified, but of course should be connected to the strategic forum/s mentioned above and links to procurers in the next section; however, given the huge potential to have influence, separate consideration is justified.

As with the first stakeholder, here there is again a pivotal action to create the *capacity* to display policy leadership. This demands a government focus and vision for construction that should be generated interactively with the work of the industry/client forum/s. This should build on ambitious aspirations for construction, extending to the alignment of activities to maximise the positive impact achieved in social, cultural, environmental as well as economic terms. By grasping policy leadership governments can provide a focus so that rules and regulations are fully informed by a broad range of stakeholders, the reputation and self-regard of the industry will be improved and attempts to look beyond 4–5 year political cycles can be instituted.

The other two action areas can then be driven by the clear policy developed, namely using public spending and legal/tax mechanisms as levers for change. Government, and the public sector more broadly, are huge customers for construction in most countries and what they spend their resources on and *how* they do so are very influential. One aspect that often arose was the potential for the public sector to provide some stability through longer term or rolling programmes of work to counter the disruptive turbulence typical of the construction sector. Legal and tax levers will impact on the orientation of players to various issues for which the direct commercial argument is not immediately apparent. Examples would be energy efficiency and the desirability of repair and refurbishment versus new build. Overall, the public sector has the potential to champion longer term and broader societal perspectives.

Procurers

These are clients in the mode of purchasers of services from the construction industry. It is at this point that much of the local context is set for given projects and so significant influence is wielded in practice. Those involved could well be influenced in their behaviour by the work of the industry/client forum/s and the example of governments' buying actions.

This is the biggest action area in terms of the amount of detail, but comes down to a simple prescription plus a range of four areas for alternative or complementary action. The simple prescription is that procurers should procure flexibly, optimally and appropriately. The implication is that this does not always happen now, but rather that often procurement is unthinkingly traditional, overly price-orientated and counter-productively induces vicious competition that releases risks which are then not well managed. The proposal is not that one approach is ideal but that each time 'win–win arrangements' should be sought in which the client's needs and expectations are clearly communicated, while other stakeholders' interests are also accommodated.

Three alternative areas for appropriate action are suggested as having particular potential to solve/resolve issues at a local level. Solutions through long-term relationships focus on stabilising relationships via 'partnering' or other permutations, so creating a shared stake in the outcomes achieved, and thus incentivising the release of added value within the relationships. This can also address issues of short-termism, limited understanding of client, lack of trust and 'organisational amnesia'. More general is the proposal for greater use of pre-qualification linked to key performance indicators. This is a way in which participants can be selected on a much wider range of factors than just price, for example 'keeping promises', but it can also enable newer factors, such as understanding of sustainability, to be built in. Pre-qualification can directly address unscrupulous low bidders, who do not subsequently deliver on their promises and may also provide a means by which the intellectual property (in the form of specialist expertise, etc.) developed by companies can be given credit.

The final area for action is not as focused as the above and concerns establishing elegant, performance-orientated processes. The emphasis is on allowing sufficient time for sound, whole-life appraisals, so releasing innovations in processes, materials and business models.

Project team

It is at this level that construction is delivered through a multitude of separate projects. Thus there is great dependence on specific procurers and other contextual factors, but also an imperative to deliver at least what was promised.

There is one main action area for the project team, and it is to work together in an integrated way, progressively managing value and risk through a series of structured events. By doing this, the delivery of the project becomes a dynamic process through which the evolution of the value delivered is kept at the forefront. The opportunity to maximise the contribution of the project is kept open, but in a controlled environment. The formality of the structuring would depend on the scale of the project, but the consistent underpinning is a focus on value and its relationship to equitable/appropriate cost. The structured events provide the opportunity to build on good relationships and integrate processes and systems. Relevant stakeholders can be involved so that shared goals across the team are maintained and, as appropriate, social, cultural, environmental and economic issues are optimally aligned. Improved information exchange and trust will result and wasteful re-design and re-work will be minimised, even in the face of the increased complexity facing many projects.

The second action area is very much subsidiary and concerns finding solutions through project information technology. Simple, accessible integrating systems, such as extranets, can have a role to play in bringing together project teams. This can exploit emerging ICT technologies, and increasingly widespread familiarity with them, to address the ineffective use of IT in construction at present.

Education and research

Education is a big contextual area, which has a number of actions that can support many of the other initiatives already covered through the avenue of longer term changes in knowledge/attitudes.

The major action area centres on clients in two senses: educating clients to be better able to perform their role; and educating those in the industry to be better able to understand clients' requirements. Thus a major driving force will be knowledgeable and focused clients, but also major impediments to be addressed are the failure of industry to understand clients and the poorly defined clients' briefs that result. This issue is of crucial importance and has to be addressed from all sides. When the role of clients is fully addressed the project will be on a firm footing and secure in the knowledge that those involved are 'doing the right things'. This then leads to the other actions where the emphasis is more on 'doing things right'.

Programmes of continuing training, education and development at all levels in the industry are needed to address a host of problem areas holding back the industry, such as: ignorance and under-valuing of management and programming practices; a reducing trade skill base; and skills/age gaps. These programmes should be infused with the client orientation covered above. A specific gap is perceived to exist in the management education of 'top management' in large construction companies. The involvement of universities and industry in a joint forum will lead to the better alignment of demand and supply in this educational area, and will help to address the view that higher education is not targeted at industry. Although it is not a major area from the workshops, research is highlighted as a potential driver of practice through innovation in processes, materials and business models and can be a route to address the improved sharing of knowledge.

Society

This is the most abstract area in one sense, but in another it highlights the practical impacts that the outcomes of construction projects have on users and wider social groupings through the built/urban environment created. The issue here is the commonplace absence of any effective voice for these groups. This appears to stem from a lack of economic power and leads to few opportunities for stakeholder involvement either in projects or institutions that could accommodate social interest groups. The

Table 19.1 Summary of major stakeholder groups and associated action areas.

Stakeholder	Action area	UK	USA	Can	Aus	Sing
Industry/client	Institutional forums for integration	■		■	■	■
	Creating integrating information frameworks		■			■
	Creating consensus 'regulatory' frameworks			■	■	
	Joint promotional themes				■	
Government	Policy leadership	■		■	■	
	Public spending as a lever	■		■		
	Legal/tax levers	■				
Procurers	Procuring flexibly/optimally and appropriately		■		■	
	Solutions through long-term relationships	■				■
	Solutions through pre-qualification/KPIs	■				
	Solutions through process characteristics	■	■			
Project team	Dynamic team/VM approach	■			■	
	Solutions through progressive process				■	
	Solutions through project information technology	■	■			
Education and research	Educating about clients		■			■
	Educating 'top management'		■			
	Continuing training and development		■	■		
	Alignment through industry/university forum		■			
	Research as a driver of practice		■	■		
Society	Hearing the user perspective		■	■		
	Hearing societal stakeholders		■	■		
	Taking the long view		■	■		

outcome is a perception that 'sharing value' is not handled equitably and perspectives, such as the operations and maintenance dimension, are under-represented. The first two action areas highlight that a better understanding is needed of the positive value of 'hearing' users and societal stakeholders through user involvement and feedback from post-occupancy evaluation, as well as working with social interest groups. If necessary this should be supported appropriately via government-led requirements.

The third and last action area goes beyond the project and concerns the long-term, strategic planning of the built environment, underpinned by an attitude that sees the built environment as an important asset that delivers social, environmental as well as economic benefits. This will be driven by practical needs, such as accommodating population growth, and the sheer necessity of making effective use of limited resources. By revealing the broader value offerings associated with construction, it is distinctly possible that the oft-quoted shortage of funds will be revealed as a mirage for a deeper felt lack of confidence in construction to deliver. This links back to all the other stakeholders and action areas.

Conclusions

So, in summary, six major stakeholder groups and a focused range of action areas have been identified, and are highlighted in Table 19.1. These are carefully selected to draw on multiple driving forces and 'attack' multiple restraining forces so maximising the systemic impact achieved.

These actions can and should address forces in all three arenas – namely: knowledge/attitudes; frameworks; and projects – although it is arguable that the fulcrum from which movement initially stems will be those actions in the 'frameworks' arena. There is a question though whether these actions are sustainable (or even meaningful) without evident success in industry practices around specific projects. Further, there is also a suggestion in the data that such initial actions may halter superficial behaviour, but that to achieve sustained progress it is crucial that knowledge and attitudes are changed on a broad front across the industry and its stakeholders.

20   Conclusions

Peter Barrett

Summary of agenda

This book has taken the 'infinity' diagram and expanded on the thinking and evidence for each of the seven parts. The discussion started with the 'holistic idea of construction' box that underpinned the scope of the 'shared vision' box. From here the discussion flowed progressively round the 'looking in' part of the diagram, from framework issues with significant levers for change around the 'balance of markets and social capital' to the more strongly project orientated aspect of 'dynamic decisions and information', leading finally to the need for 'evolving knowledge and attitudes' to achieve long-term, internalised change in the industry. This then led to the 'looking out' part of the diagram through the link of the shared vision to an 'awareness of the systemic contribution' of construction, which can then support the active 'promotion of the full value delivered to society'. The various issues surfaced in the sections above are summarised diagrammatically in Figure 20.1.

It is worthwhile distinguishing the nature of the two halves of the diagram. 'Looking in' stresses actions to enhance the performance of the industry. As such, it deals with complex issues for which simple, uni-dimensional solutions are not available. Thus, the signature of this half is 'appropriateness', or balance, between various factors. What this book aims to do is pinpoint the main aspects around which the issues in each area appear to rotate. However, the argument is categorically *not* that long-term relationships should replace contracts; that all information should by kept and recycled; or that education and research alone can solve the problems of the industry. Centrally, it is also not arguing for a single monolithic vision for the industry, but rather a vibrant dialogue with energised partners. Some level of coherence and focus is, however, desirable and the schema set out in this book is a suggested landscape within which such a debate and the consequential actions can fruitfully be pursued.

The 'looking out' half of the diagram is different in emphasis and broadly speaking is underpinned by an argument for looking broadly beyond existing categorisations, mindsets and images. This is evident in the 'holistic' area where existing economic conventions are problematic; in the 'systemic' area where the accounting of contributions from construction is lacking; and in the 'promotion' area where the industry is significantly undervalued, by others and by itself.

Overall, these elements make up a vision of how construction can and should move coherently on seven linked aspects. Providing that performance improvements broadly match positive changes in perception, then a virtuous cycle can be expected to operate. However, a concerted effort over a number of years or even decades will be needed to achieve significant and enduring change.

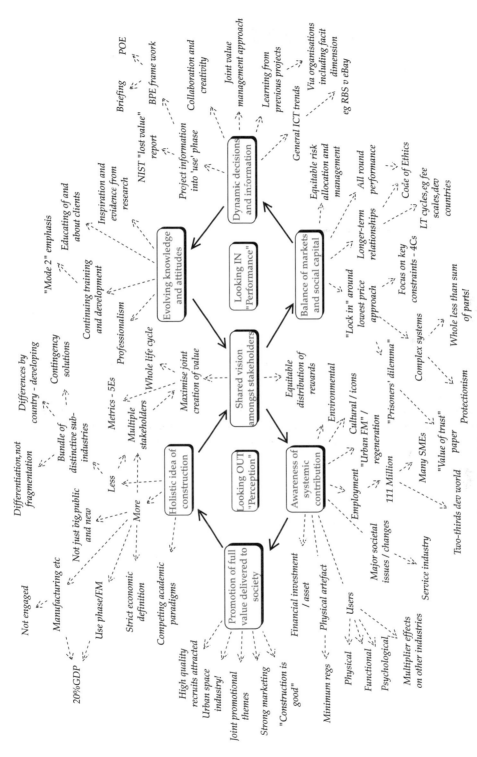

Figure 20.1 Summary of issues connected to 'infinity' diagram.

Table 20.1 Summary of impacts and research actions.

Vision	A construction industry that is highly valued by society						
Society impacts	*Owing to an appreciation of the:*						
	Excitement attached to the achievements of, and opportunities within, construction	Multiple *contributions* (short and long term direct and indirect) that result from construction	*Scale* and reach of construction industry within the economy	Sense of *direction* and aspiration shared within the construction industry	*Professional* and ethical approach in client and other relationships	Flexible, thoughtful and *sophisticated* approach to ongoing problem-solving	Strong *knowledge* base and *skill* sets built on progressive education
Revaluing construction priorities	*Driven by:*						
	Looking OUT ⟶			Shared vision amongst stakeholders	⟶		Looking IN
	Promotion of full value delivered to society	Awareness of systemic contribution	Holistic idea of construction		Balance of markets and social capital	Dynamic decisions and information	Evolving knowledge and attitudes
Research actions (examples)	*Supported by research that provides:*						
	Case studies of *exemplary* (4Cs) projects and initiatives, illustrating benefits to all stakeholders involved	Frameworks and measurement techniques to *account* for multiple hard and soft contributions	Powerful arguments and *advocacy* to drive a reassessment of economic/conceptual definitions	A basis for active engagement with, and contribution to, the *strategic debate* and implementation strategies	Increased conceptual and practical understanding of how (and why) to *move past 'least cost'* with confidence	The concepts and practical tools to address *interface barriers* to the flow of explicit and tacit information	Concepts, evidence, and *enthusiasm* to infuse the education/training provision to students and practitioners

This book has covered a lot of ground. The main contributions are:

- A drawing together of the results of surveys and workshops in five countries as well as inputs from a variety of workshops and from the project's scientific panel.
- A definition of 'revaluing construction' as the maximisation of the value jointly created by the stakeholders to construction and the equitable distribution of the resulting rewards.
- A relatively simple overview of the seven major factors to be taken into account if the above aspiration is to be achieved, in the form of the 'infinity' diagram.
- A discussion of each of the seven areas linking with relevant material from a range of specialist contributors.
- A way of envisioning 'revalued' projects in terms of exemplars analysed around the '4Cs' model.
- Proposed action areas for specific stakeholders, highlighting the uneven leverage available and proposing a sequence of events that can maximise the influence available.

Although many research findings have been used, the role of research has not been unduly emphasised in the transformation sought. Table 20.1 is the last substantive contribution and endeavours to both indicate the anticipated impacts of actions in the seven areas within society generally and the role research should have in moving things forward.

Researchers can create exemplary cases, improve the measurement of benefits, re-define the economic models used and inform strategic debates. They can also provide concepts and tools to go beyond 'least cost' and to address barriers to information flows and, of course, enthusiastically contribute to education and training. It can be hoped that as a result 'society' will appreciate the excitement of construction, its multiple contributions, sheer scale, sense of aspiration, professionalism, sophistication and strong knowledge/skill base. However, the major focus of commitment and effort must come from within the industry itself as it moves out of a defensive mode to confidently take forward its pivotal role in society.

Appendices

Appendix 1   Members of the CIB Revaluing Construction Steering Panel

Robert Amor   University of Auckland, New Zealand
George Ang   PSIBouw, The Netherlands
Peter Barrett   University of Salford, UK (Chair)
Stephen Bennett   Liverpool John Moores University, UK
Martin Betts   Queensland University of Technology, Australia
Luc Bourdeau   CSTB, France
Roger Courtney   UK (an independent consultant)
Ron Craig   Loughborough University, UK
Gerard Devalence   University of Technology Sydney, Australia
Richard Fellows   Hong Kong Polytechnic University, China SAR – Hong Kong
Peter Fenn   University of Manchester, UK
Greg Foliente   CSIRO, Australia
Jimmie Hinze   University of Florida, USA
Stephen Kendall   Ball State University, USA
Matti Prins   Delft University of Technology, The Netherlands
Steve Rowlinson   Hong Kong University, China SAR – Hong Kong
Les Ruddock   University of Salford, UK
Danny Then   Hong Kong Polytechnic University, China SAR – Hong Kong
   Shiem-Shin
Wilco Tijhuis   University of Twente, The Netherlands

Appendix 2 Sequence of activities and sources feeding into the CIB revaluing construction proactive theme

Event/activity	Ideas
CIB/UMIST 2000–2002 survey with 223 responses (60% non-research) and workshops with 59 attendees in three locations (Europe, North America and Asia/Pacific) Source: CIB Information Bulletin, Feb. 2002	*Questionnaire* found that: **All** re-eng themes put forward thought to be on average more than 'quite important', with no single theme of clearly leading importance Over 80% considered that the underlying issues in their country would also be found in others 45% of industry respondents would be prepared to devote more than 3 days pa to the issues (and more for researchers) *Workshops* found that: Reinforced broad consensus on issues and direction needed Some differences in priorities by region were revealed The co-ordinators of CIB groups were also polled and a locus around W65, 78 and 92 was revealed for bottom-up work *Keywords*: more efficient, integrated and attractive; better service, and product; changes in management structures, relationships and attitudes, mostly focused people or 'culture' with technology underpinning, not driving (although new prefab/industrialised focus recommended to fill gap). Clients a key influence via procurement

Event/activity	Ideas
	Value links hard and soft aspects of construction – proposal to move from 'Re-eng' to 'Revaluing construction', David Hall's (of BAA) idea at workshop
Meeting of CIB Co-ordinators in CUBE, Manchester, 4th Sept. 2002	Six co-ordinators met and it was agreed that: A working definition of Reval should be developed as the ideas are firmed up. Reval usefully loose and broad concept
	It was not intended to exclude dev countries, although emphasis was on developed at present
	Resonance was evident with the work of W78, 103, 92, 65 and 89 (education) and TG47 and 41
	PSB possible structure of: shared vision/trajectory, models, metrics, change processes/management and tools/techniques
	Stress on *need* for good conceptual models of construction that will be viewed from multiple Ws
	Martin Betts presented the W78 research agenda which stressed seven key themes: product/process *model*, not document driven; seamless life-cycle for information; use past knowledge; internet procurement; visualisation through life-cycle; better simulation/what-if; and change management/process improvement
Meeting of CIB Co-ordinators in Cincinnati, 11th Nov. 2003	Open meeting of around 50 with many specific contributions: W55 – economic data straightforward on shape of industry, but as several responded, misleading re full scope
	TG31's meso-level view of construction interesting re full scope and bundles of industry notion
	Importance of non-economic aspects of value (social, cultural), role of sub-contractors, *product suppliers*, services dimension, re-use of existing buildings, multiple stakeholders, buildings as assets that create value. (Idea to artefact; artefact to asset)
	Need for metrics to understand where we are and improvement achieved – American Society of Value Engineers have definitions for many aspects
	Emphasis on means here not aims

(Continued)

Event/activity	Ideas
	Some *keywords* from this meeting: coalitions, service industry, buildings *in use*, capacity, solutions, complex systems, longer term, learning
Request to co-ordinators for definitions after Cincinnati	*Scope of construction* via stakeholders notion
	Get to all those activities that deliver changes in the built environment to owners, users and society in general – i.e. incl. FM!
	Contains notion that construction is in fact a collection of sub-industries, with different supply and demand players
	Given wide scope and diversity of interests need definition/vision/model/s to integrate
	Keywords of definition: improving, value flows, hard and soft, stakeholders, products and services, change and built and human environments
	Cognitive map summary of WCs/TGs and Reval
Reval conference, Manchester, 3–4 Feb. 2003	Mainly industry and confirmed interest in various issues
	Need for breaking out concept of value and working through implementation issues in different policy contexts. PSB sent ppt for MB, which raised dichotomy of academics focused on single issues across many sub-sectors and industry the converse
BRI 31(2) *Special Reval Issue*, March 2003	Summarised UMIST work and PSB workshops, RC and GW proposing three main aspects: Better understanding needed of how constructed assets add value for financiers and users
	Better capture of value through project life-cycle in terms of profits to companies and learning to people
	Based on above two – a re-valuation of the image of the industry
	Issue included other papers with conflicting views: Koskela presents a new conception of construction arguing *more than structural change* is needed and more effort is needed at the level of operations
	Green & May arguing that an overemphasis on a customer focus depletes the industry's capacity to actually deliver
	Winch arguing that mass or *lean production models are inappropriate* for construction as construction not suitable

Event/activity	Ideas
	Ballard & Howell 'demonstrating' that the lean production approach is superior to traditional practices
	Barlow et al. arguing that mass customisation, as in Japan, may be a paradigm for house building but *not other market segments*
	Gibb & Isack showing that *clients do not confirm* that benefits of pre-assembly are forthcoming
	Arbulu et al. pushing the value stream analysis technique whilst acknowledging that *re-eng in construction has had mixed results*
PSB *Programme of Activities* proposed to create 'Agenda 21 for Revaluing Construction', Sept. 2003	Programme to include focused events to draw CIB groupings in, elements in already planned major CIB events and a questionnaire and workshops in five countries overseen by a scientific panel of co-ordinators and supported by selected authors and a network of academic institutions. Completion planned for Helsinki, June 2005
Report to CIB Board up to Singapore Oct. 2003/Toronto June 2004	Much as above elements in Manchester, Cincinnati, Singapore
	Issues to reinforce here are: Central issues from cognitive map: product and process models, structure of the industry, visions of construction, benchmarking, indicators of better value, process from normal to new practices, life-cycle information, e-procurement, visualisation/VR, change management, company-based learning, procurement as a driver for change and the role of culture and trust
	W92 and value alignment approach; TG 31 and meso-industry model re sub-sectors
	Final report will aim to clearly articulate for major stakeholder groups: vision, models, success indicators, flexible *5–10 year* recommendations and contextualised advice on tools and techniques to support change efforts
Meeting of CIB Co-ordinators in Singapore, 23 Oct. 2003	Circa 15 co-ordinators/members met and agreed that: Construction *is a set of sub-industries*, it cannot be treated as a homogenous whole
	Moreover, drivers for change in different countries will vary and so a *global recipe would not be appropriate*, however, there are common issues

(Continued)

Event/activity	Ideas
	Focusing the work around how these *sub-industries* satisfy their clients where they are demanding improvement, including operations as well as building
	Value goes beyond financial and should include: value to supply interests, including the *workforce and the community*
	Important that the *multiplier effect of construction* on other sectors is highlighted especially for many governments who perceived construction as un-dynamic compared to these sectors
	A possible way forward would be to select a sub-sector where client dissatisfaction is evident and work through the value held by different parties
PSIBouw visits to Norway and Denmark 24–27th Feb. 2004 General note	Plenty of input from consultants and Govt, but contractors view muted
	Can't see initiatives as 'demonstrator projects' as have to comprehensively treat the system as a whole to *escape 'lock- in'* (mentioned in Nor and DK with *awkward* implications of 'ring-mastering' things in face of free market emphasis), incl. aspects such as: *Creating forums* for under-represented groups and to focus disparate demands, e.g. clients' forum and liaison group between Govt Ministries in Denmark
	Vertical strategic connections through *selective communication*, e.g. UK and same Norway
	Symbolic actions such as joint Code of Ethics in Norway (2003)
	Extensive programme of facilitated workshops to achieve *ownership* of ideas, e.g. in Norway where academic report not taken up until did this. (DK clients' forum similar mode)
	So long-term change mg approach important – potential of *Kotter's model and possible eg for NL*
	Care needed with 'nice' fashionable approaches such as *partnering/Samspill. Look beyond names* – often still underpinned by trad. contracts, but main variable seems to be % of procurement decision based on price (although two stage can mean second stage mostly price)

Event/activity	Ideas
	Need care re assumption that public sector leads private when vice versa seems more likely in terms of getting good deal – but publ. W more prominent as national initiatives. EU tendering regs used as reason, but not as restrictive as thought. Big public sector projects *atypical and should not be taken as a template for the sector*
	Facilities management becoming more prominent (Nor/DK)
	Idea of creating vertical *integration at three levels* in terms of *theory* (incl. use-phase), *policy* (incl. client and FM groupings) and in *practice* via partnering, etc.
PSIBouw DK specifics Meeting note	Major long term initiatives – some have worked very well, e.g. recycling and energy, where backed by taxes, etc. Others have dissipated after initiative finished, e.g. system building for social housing. This is some evidence that *cannot really manage the market*. The most successful initiative aimed at only improving traditional approach, *not radical*
	Have *general requirement* (law) to take *LCC* considerations into account and to assess projects for *PPP* potential benefits beyond initial 'problem' good to maintain momentum, e.g. DK expertise in recycling now an export strength
	Taskforce 2000 Report identified *four drivers* for change:
	more prof *clients*, greater *competition* in industry, improved *cooperation* in construction projects and more efficient use of *knowledge*
	Push on comprehensive benchmarking via Danish Building Evaluation Centre. Set up Feb. 2001 building on UK M4I and DK *'From Tradition to Innovation'* 2001 report
PSIBouw Norway specifics Meeting note	Danish Association of Construction Clients seems to be an exemplar of how to *incrementally set up a lean forum* – only professional clients with no supply-side involvement
	General initiatives over many years show progression from academic studies to leading industry reports to general engagement via mixed workshops – major outcome *understanding of 'lock-in'* and need to work together to move forward (PSB link Prisoners' Dilemma via Sacks)
	New hospital in Trondeim where *psychologist* used to assess teams as major part of selection

(Continued)

Event/activity	Ideas
	RIF annual survey found that *SMEs <10 staff were the most profitable* and some larger companies loss-making – maybe this explains why not so motivated to be engaged in these big initiatives
	Long-term cycle of removing fee scales – fierce competition – problems – move to compete on quality, etc. not just price!
CIB World Congress Toronto, 2–7 May 2004, Various Reval elements	PSB keynote – ppt slides introduced notions of: 'impressive examples of construction', vision/defin. as above, holistic/contingency models, model of ± value adding and non-v. adding through life-cycle, gaps in current research identified – namely issues of: people/quality of life, 'urban FM', PBB, managing information, IT application, etc., notion of looking into construction and project company nexus, plus idea of putting in and getting out value, bundle of industries illustrated with Mopmit, Galbraith used for structure plus other solutions, work with highly diff. ind. giving value back to stakeholders to be sustainable (ind. doing OK – is good), benchmarking performance – vertical scientific v 'expedia/trip advisor', e.g. of RBS IT bidding like ebay; looking out from construction – *SCRI vision* (Sci Panel v keen), radically changing world (Hamish McRae) re construction/BE implications; searching for 'improvement paths' using *5Es* allocated across *five stakeholders*.
	Workshop 1: Revaluing the agenda – focused on the development of construction management research around papers by Roberto Pietroforte (tracking papers in two major journals over 20 years – trend towards con mg from eng base) and Pekka Houvinen (journal papers over 12 years assessed against 8 mg approaches – pointing to complex systems' approach linking academic fields of: con eco and mg, real est dev, proj mg and ind mg – national industries as viable systems) Quote from 'UK Govt': 'I don't want to understand construction – I want to change it'. Main feeling from meeting of *competing academic paradigms that seem incommensurable at present*.
	Workshop 2: Revaluing construction practice, based on papers by Matt Prins (science war between arch, tech and mg – arguing for role of *creativity and design and importance of architectural value*) and Alistair Gibb (on use of IT by SMEs – need for IT use to make sense in strategy of firms)

Event/activity	Ideas
	Meeting of scientific panel – liked presentation especially SCRI vision. Not keen on aligning value as misses tension and complementarities aspect. Liked 'value in' and 'out', project/company interface and life-cycle notions and saw potential in 5Es criteria. Liked emphasis on value in use and accepted bundle of industries argument (new v repair and maint'?).

Abbreviations: arch, architecture; con, construction; def, definition; dev, developing/development; DK, Denmark; eco, economics; ind, industry; maint, maintenance; mg, management; NL, The Netherlands; Nor, Norway; ppt, Powerpoint; prj, project; pub, publication; tech, technical; v, value/versus.

Appendix 3   Sample revaluing construction covering letter and survey questionnaire

Dear Colleague,

Re: Revaluing Construction: An International Study

Around the world, construction processes and practices are under scrutiny. Changing markets, new technology and rising client expectations are stimulating radical reviews of how the industry can be re-engineered to enhance its performance.

Your company has been identified as one that will have an important perspective on construction performance. Therefore, I am writing to you for your help in undertaking our research. The purpose of this questionnaire is to examine your views and beliefs about the appropriate *value* derived from the construction industry for its key stakeholders. It is intended to identify current practice and future trends of stakeholder needs in Australia, Canada, Singapore, United Kingdom and the United States of America. This study is being undertaken by the University of Salford's 6* rated Research Institute for the Built and Human Environment (the highest ranked in the UK), on behalf of the International Council for Research and Innovation in Building and Construction (CIB), the leading international network for innovation in construction. For the British survey, we are collaborating with Dr... at the Salford Centre for Research and Innovation (SCRI, University of Salford, who will be happy to confirm the importance of this study.

Your responses are confidential. Subsequent reports as a result of this questionnaire will be written in such a way that no individual or organisation can be identified. Please answer all questions giving your first and natural answer, and base your answers on your experience from an organisational and industry perspective. In this way, you should be able to complete this questionnaire in approximately 10 minutes. Space is provided at the end of the questionnaire if you would like to expand on your answers.

If you have any queries, please do not hesitate to get in touch with me or Dr .... All respondents who identify themselves will be sent a copy of the results. A follow-up workshop will be held in Salford on the 15th June; please indicate on the questionnaire if you would like to attend this event. Please return the completed questionnaire by

Friday 14th May 2004 by facsimile or post using the return envelope provided. Your views will have an impact on the strategic development of construction worldwide.

Thank you for completing this questionnaire.

With kind regards,

Professor Peter Barrett

PS: Although every effort has been made to target the most appropriate person within your organisation, if you believe that somebody else, in addition or instead of you, should complete the questionnaire please feel free to copy it or pass it on

Revaluing Construction: An International Study

Name: *(optional)*
Company name: *(optional)*
Nature of business: *(please circle)* Client   Design team   Project Management
                                   Contractor   Sub-contractor   Supplier/manufacturer
Approx. number of employees:
Position held/ designation:
Address: *(optional)*
Email: *(optional)*
Telephone: *(optional)*
Facsimile: *(optional)*
I cannot/would like to attend the follow-up workshop *(please delete)*. Please send further details to the address above.

Please indicate your level of agreement or disagreement to the following statements by circling the appropriate number.

	Strongly disagree	Disagree	Neutral	Agree	Strongly agree
Q1 The following stakeholders are **very important** contributors to the value provided by construction:					
- clients	1	2	3	4	5
- end-users	1	2	3	4	5
- design team	1	2	3	4	5
- project manager	1	2	3	4	5
- contractors	1	2	3	4	5
- sub contractors	1	2	3	4	5
- suppliers/manufacturers	1	2	3	4	5
- society at large	1	2	3	4	5

	Strongly disagree	Disagree	Neutral	Agree	Strongly agree
Q2 The value gained from construction is **very high** for the following stakeholders:					
- clients	1	2	3	4	5
- end-users	1	2	3	4	5
- project manager	1	2	3	4	5
- design team	1	2	3	4	5
- contractors	1	2	3	4	5
- sub contractors	1	2	3	4	5
- suppliers/manufacturers	1	2	3	4	5
- society at large	1	2	3	4	5
Q3 The following factors have a **very significant** influence on the value gained by stakeholders from construction:					
- existing procurement routes / contractual relationships	1	2	3	4	5
- traditional structure of the construction industry	1	2	3	4	5
- role of public policies/institutional bodies	1	2	3	4	5
- use of information technologies	1	2	3	4	5
- alignment of stakeholder objectives	1	2	3	4	5
- predominance of small/medium-sized organisations	1	2	3	4	5
- communication between project participants	1	2	3	4	5
- other: *(please state)*	1	2	3	4	5
Q4 I am **very convinced** that the following factors enable stakeholders to gain appropriate value from their involvement in construction:					
- existing procurement routes/contractual relationships	1	2	3	4	5
- traditional structure of the construction industry	1	2	3	4	5
- role of public policies / institutional bodies	1	2	3	4	5
- use of information technologies	1	2	3	4	5
- communication between project team participants	1	2	3	4	5
- predominance of small/medium-sized organisations	1	2	3	4	5
- alignment of stakeholder objectives	1	2	3	4	5
- other: *(please state)*	1	2	3	4	5

	Strongly disagree	Disagree	Neutral	Agree	Strongly agree
	1	2	3	4	5

Q5 My organisation is generally **much** too slow to embrace new practices and procedures that could improve construction performance: — 1 | 2 | 3 | 4 | 5

Section 3: Industry perspective

Q6 The following factors **significantly** inhibit the widespread adoption of new practices that could improve construction performance:

	Strongly disagree	Disagree	Neutral	Agree	Strongly agree
- lack of government role model	1	2	3	4	5
- institutional bodies protecting their own interests	1	2	3	4	5
- industry/client protecting their own interests	1	2	3	4	5
- inappropriate legislation	1	2	3	4	5
- strict health & safety requirements	1	2	3	4	5
- professional indemnity insurance cover	1	2	3	4	5
- lack of research and development	1	2	3	4	5
- lack of awareness of improvement initiatives	1	2	3	4	5
- lack of education and training	1	2	3	4	5
- lack of technical innovation/implementation	1	2	3	4	5
- lack of innovative procedures/practices	1	2	3	4	5
- lack of clear benefits	1	2	3	4	5
- extensive inter-organisational change required	1	2	3	4	5
- temporary nature of construction projects	1	2	3	4	5
- belief that the industry is doing well without innovation	1	2	3	4	5
- other: *(please state)*	1	2	3	4	5

Q7 The following are effective drivers for **significantly** increasing the value delivered by construction:

	Strongly disagree	Disagree	Neutral	Agree	Strongly agree
- new procurement methods	1	2	3	4	5
- integrated/interoperable IT tools	1	2	3	4	5
- new management techniques	1	2	3	4	5
- socially responsible design	1	2	3	4	5
- environmental awareness	1	2	3	4	5
- life cycle value/value–price leverage	1	2	3	4	5
- an emphasis on culture/trust issues	1	2	3	4	5

	Strongly disagree	Disagree	Neutral	Agree	Strongly agree
- continuous learning from experience/past projects	1	2	3	4	5
- equitable allocation of risk	1	2	3	4	5
- other: *(please state)*	1	2	3	4	5

Q8 The construction industry is currently **radically** improving its performance in the following areas:

	Strongly disagree	Disagree	Neutral	Agree	Strongly agree
- on-time delivery	1	2	3	4	5
- value for money	1	2	3	4	5
- quality	1	2	3	4	5
- safety record	1	2	3	4	5
- public image	1	2	3	4	5
- research & development	1	2	3	4	5
- customer relations	1	2	3	4	5
- education & training	1	2	3	4	5
- other: *(please state)*	1	2	3	4	5
construction industry in general	1	2	3	4	5

Section 4: Additional comments

Thank you for your co-operation. Please return completed questionnaires by **Friday 14th May 2004** to:

Dr...
Facsimile: +44...*or* email: ...

University of Salford
Salford Centre for Research and Innovation (SCRI)
Research Institute for the Built and Human Environment
Bridgewater Building
Meadow Road
Salford
M7 1NU
United Kingdom

Appendix 4   International survey questionnaire results

1.0 Revaluing construction: an international study

Around the world, construction processes and practices are under scrutiny. Changing markets, new technology and rising client expectations are stimulating radical reviews of how the industry can be re-engineered to enhance its performance.

An international survey questionnaire was administered in May 2004 to examine the *value* derived from the construction industry for its key stakeholders. It was intended to identify current practice and future trends of stakeholder needs in Australia, Canada, Singapore, the United Kingdom and the United States of America. The study was undertaken by the University of Salford's 6* rated Research Institute for the Built and Human Environment (the highest ranked in the UK), on behalf of the International Council for Research and Innovation in Building and Construction (CIB), the leading international network for innovation in construction. Collaborating institutions included:

- Queensland University of Technology (QUT), Australia
- University of Montreal, Canada
- National University of Singapore (NUS), Singapore
- Massachusetts Institute of Technology (MIT), USA

It is anticipated that the survey findings will be validated and consolidated through a series of international workshops in each survey country, the results of which will be published worldwide and used to inform leading research agendas.

1.1 Survey sample

The questionnaire was sent to 200 organisations in each survey country: Australia, Canada, Singapore, UK and USA. In order to capture an internal and external perspective of construction, the survey sample included organisations directly within the industry, such as contractors and architects, and also those who are usually considered as indirectly involved, such as clients and suppliers:

- Contractors (sub-contractors, engineers, etc., 50 organisations selected)
- Design team (architects, engineers, surveyors, project managers, etc.; 50 organisations selected)
- Manufacturers/suppliers (50 organisations selected)
- Clients/end-users (50 organisations selected)

The spectrum of disciplines was selected to reduce the chances of creating a survey questionnaire that would be discipline-biased and thus limit the generalisability of the results. Furthermore, this also reduced any chances of any specific product bias (i.e. building type) from the desired generically applicable results. The organisations were selected at random from various professional institutional lists of members, and included both small and large firms. Table A4.1 illustrates the sources that were used to nominate the sample.

1.2 Development of survey questionnaire

The revaluing construction issues listed in the survey questionnaire were identified through various CIB Working Group and Task Group meetings. A five-point balanced Likert scale (1 = strongly disagree, 2 = disagree, 3 = neutral, 4 = agree or 5 = strongly agree), commonly used in attitude assessment, was adopted to ascertain the respondent's level of agreement or disagreement to statements related to construction in their country, both at organisation and national levels. The questions were divided into four sections:

- Background information
- Organisational perspective
- Industry perspective
- Additional information (space for further comments)

A covering letter was sent with each survey questionnaire, which stated its purpose and indicated that all the responses would be confidential and no individuals would be identified in the subsequent reports. In addition, instructions on how to complete the questionnaire were included at the head of each section. A copy of the questionnaire and covering letter is given in Appendix 3.

In order to test the validity of the questionnaire, it was sent to the collaborating academic institution in each survey country for comments on its applicability, and also to eight practitioners in the UK. This was to check for clarity, terminology and consistency of the questions/topics covered, from both international and industrial perspectives. The recommended changes were made to the survey questionnaire. The study was finally sent to the 1000 nominated organisations, followed by two further follow-up chases for responses. Although a return self-addressed envelope was included with the survey, postage was not pre-paid.

Table A4.1 Sampling frame for study.

Country	Clients	Contractors	Design Team	Manufacturers
Australia	Property Council of Australia http://www.propertyoz.com.au	Civil Contractors Federation http://www.ccf.org.au	The Royal Australian Institute of Architects http://www.raia.com.au	Australian Building and Construction Links http://www.spec-net.com.au/company/linka us.htm
	Housing Industry Association http://www.hia.asn.au	Master Builders Association http://www.mbasa.com.au	Consulting Surveyors Australia http://www.surveying.org.au	Building Industry Specialist Contractors Organisation http://www.amca.com.au
Canada	Canadian Home Builders Association http://www.chba.ca/	Listings Canada http://www.listingsca.com/Business/Construction	Canadian Institute of Quantity Surveyors http://www.ciqs.org/	Listings Canada http://www.listingsca.com/Business/Construction
			Royal Architectural Institute of Canada http://www.raic.org/	
Singapore	Supplied by NUS, Redas www.redas.com.sg	Supplied by NUS, Building & Construction Authority www.bca.gov.sg	Supplied by NUS, Singapore Institute of Architects www.sia.org.sg	Supplied by NUS,

Country					
UK	Confederation of Construction Clients http://www.bath.ac.uk/management/agile/introduction/CCC.html SCRI Forum members	Construction News http://www.cnplus.co.uk/ SCRI Forum members	Singapore Institute of Surveyors www.sisv.org.sg	The Architects Journal http://www.ajspecification.com/	UK Construction http://www.ukconstruction.com/
	Design & Build Foundation SCRI Forum members		SCRI Forum members	Royal Institution of Chartered Surveyors http://www.rics.org.uk SCRI Forum members	The Architects Journal http://www.ajspecification.com/ SCRI Forum members
USA	USA Building Guide http://www.misronet.com/usaconlist.htm	USA Business Directory http://www.usalinker.com/	USA Business Directory http://www.usalinker.com/	USA Business Directory http://www.usalinker.com/	
		Construction Business Directory http://www.bizeurope.com/bsr/csr/construc.htm	Construction Business Directory http://www.bizeurope.com/bsr/csr/construc.htm	Construction Business Directory http://www.bizeurope.com/bsr/csr/construc.htm	
		USA Building Guide http://www.misronet.com/usaconlist.htm	American Institute of Architects http://www.architecture.com/go/Architecture/Reference/Links.1135.htm		

2.0 Summary of results

The data from the survey questionnaire was entered into Microsoft Excel for analysis. Using the software, descriptive analysis was compiled for every question. This incorporated a frequency count of the possible responses, the percentages, mean, mode, median, standard deviation, standard error, range, confidence level and sample variance. This choice of statistical analysis was selected as it was deemed appropriate for ease of relaying back the findings to academics and industrialists alike. The results are summarised below, and are illustrated using pie charts, scatter graphs, tables and line graphs. The axes of the graphs are labelled according to the same Likert scale that was used in the survey questionnaire.

2.1 Response rate to survey questionnaire

Table A4.2 illustrates the number and percentage of the sample population who were sent, and subsequently responded to, the survey questionnaire by country and by stakeholder. In total, there were 214 (21.4%) respondents of which only 190 (19%) were usable.

Figures A4.1 and A4.2 illustrate the proportion of responses by country and by discipline. The response to the American survey (9%) is particularly low, and therefore the results must be viewed in context. The lower number of responses from Australia, Canada, Singapore and USA than the UK (see Figure A4.1) may largely be due to the fact that this is an international study and the self-addressed envelope was not pre-paid, the addresses of the organisations selected at random were no longer in use and/or the questionnaire was inappropriate for their field of work. There was a greater response rate from the stakeholder groups that are directly involved in the construction industry (contractors and designers) than those who are generally considered to be indirectly involved (manufacturer/suppliers and clients/end-users; Figure A4.2). Perhaps more work should be undertaken to target and include these stakeholders. However, at least on the face of it, it would appear that these stakeholders consider 'revaluing construction' to be less their responsibility than do contractors and designers. Of the respondents, 23% were from micro-sized enterprises (organisations with between 0 and 9 employees), 26% were from small-sized enterprises (organisations with between 10 and 99 employees), 38% were from medium-sized enterprises (organisations with between 100 and 499 employees) and 13% were from large-sized enterprises (organisations with over 500 employees).

As the questionnaire was followed-up twice, it is possible to compare the results with the 'late responders,' i.e. those who returned questionnaires in the last round, with the rest. This is relevant, as it is known, in general, that late responders are like non-responders. In this case the late responders were found not to be markedly different from the main response, thus it is reasonable to infer that the responders broadly describe the whole population.

It was similarly found that, in general, responses from the final countries were very similar and so can be taken as a whole. This mitigates the low response from some countries. Where marked differences are apparent these are highlighted.

Table A4.2 Response rates to survey questionnaire by stakeholder and country.

Discipline	Number sent to each country	Australia		Canada		Singapore		UK		USA		Total	
		No. of respondents	% of given subset	No. of respondents	% of given subset	No. of respondents	% of given subset	No. of respondents	% of given subset	No. of respondents	% of given subset	No. of respondents	% of given subset
Contractors	50	9	18	11	22	16	32	34	68	9	18	79	31.6
Design team	50	14	28	10	20	12	24	32	64	9	18	77	30.8
Manufacturer/supplier	50	4	8	3	6	1	2	5	10	0	0	13	5.2
Client/end-user	50	1	2	1	2	9	18	10	20	0	0	21	8.4
Total	200	28	14	25	12.5	38	19	81	40.5	18	9	190	19

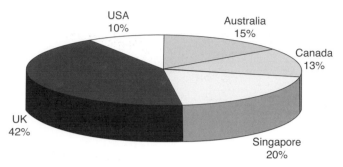

Figure A4.1   Questionnaire respondents by country.

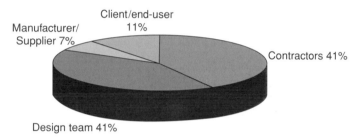

Figure A4.2   Questionnaire respondents by discipline.

2.2 Response to '(the following) stakeholders are *very important* contributors to the value provided by construction' and 'the value gained from construction is *very high* for (the following) stakeholders'

Respondents were invited to indicate the value provided and gained from construction in respect of the key stakeholder groups; the results are illustrated in Figure A4.3. The mean response fell between the 'neutral' and 'strongly agree' criterion. It is clear from the scatter graph that all stakeholder groups felt that they did make very important contributions and, in return, gained very high value from construction. The clients, end-users and 'society at large' groups gained more from construction than they contributed.

The fact that all responses were, on average, positive is interesting in itself given the demanding formulation of the questions.

The value provided and gained by construction in each survey country followed a similar pattern (see Figures A4.4–8), all results fell between the 'neutral' and 'strongly agree' criteria. The only anomalous result was Australia, whereby the 'society at large' group was foreseen as gaining but not providing value. Notably, the Australians believed that (Figure A4.4) clients and end-users gained and provided value from construction in equal amounts. More too, the project manager gained more value from construction from what they originally put in!

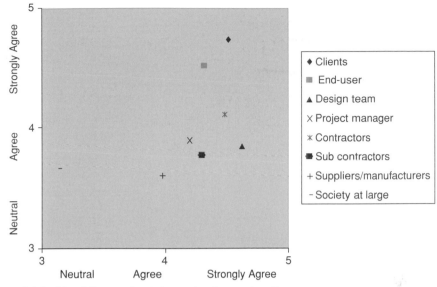

Figure A4.3   Providing and gaining value in construction.

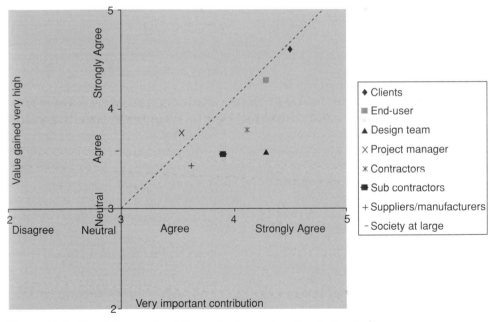

Figure A4.4   Providing and gaining value in construction in Australia.

The Canadians (Figure A4.5) felt that clients/end-users, society at large and manufacturers/suppliers were gaining more value from construction from what they originally put in. The other remaining stakeholder groups were more closely clustered in

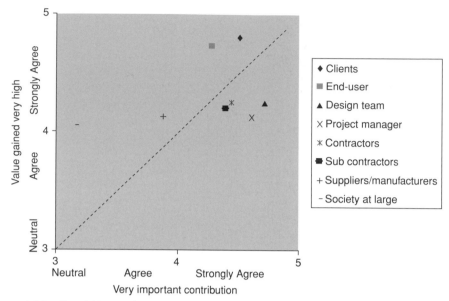

Figure A4.5   Providing and gaining value in construction in Canada.

response than the other survey countries – they provided more value to construction than what they gained back.

The Singaporean response (Figure A4.6) best mirrored the mean responses of the whole survey (Figure A4.3).

The British response (Figure A4.7) also reflected the result of the mean response of the whole survey (Figure A4.3). However, the clients seemingly provided more value to construction than what they received. This may be due to the fact that large/experienced clients in the UK have been, to a certain extent, driving the construction industry forward. For example, Boots, British Airports Authority, London Underground and Sainsbury's have developed their own process approach to construction, adopted partnering arrangements and have started to dictate their practices across all their construction projects. The Construction Round Table was established in the late 1990s to cater for this growing trend of client empowerment.

The American response (Figure A4.8) emulates the same pattern as Singapore (Figure A4.6) and the mean response of the whole survey (Figure A4.3).

2.3 Response to '(the following) factors have a *very significant* influence on the value gained by stakeholders from construction' and 'I am *very convinced* that (the following) factors enable stakeholders to gain appropriate value from their involvement in construction'

Respondents were asked to indicate their agreement or disagreement to the relative significance of a number of key factors in terms of influencing and enabling stakeholder value in construction. The results are shown in Figure A4.9. All the factors questioned

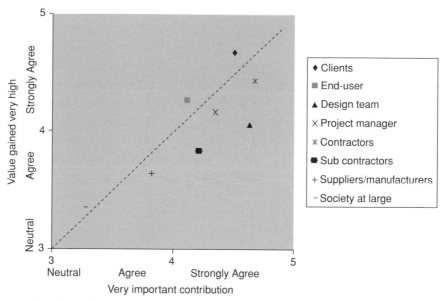

Figure A4.6 Providing and gaining value in construction in Singapore.

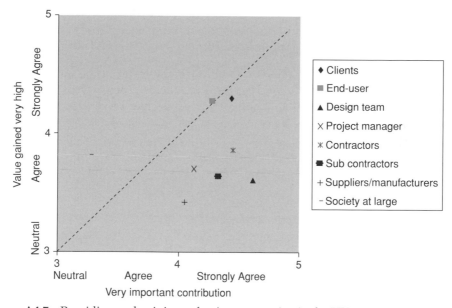

Figure A4.7 Providing and gaining value in construction in the UK.

were considered to influence and enable stakeholder value – all the responses fell between the 'neutral' and 'strongly agree' criteria. The categories regarding the use of information technologies, communication between project team members and the predominance of small/medium-sized organisations are most fitting, in that they both influence and enable value in equal amounts. It can be seen that a group of factors

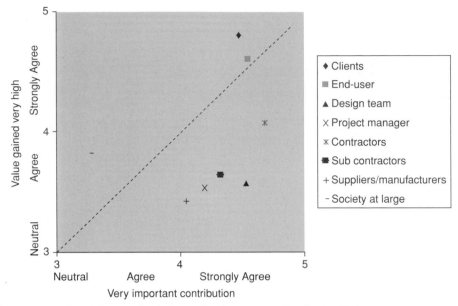

Figure A4.8   Providing and gaining value in construction in the USA.

fall below the line, indicating that they are rated more strongly for influence on value than for their impact on the appropriate allocation of value to stakeholders. These would appear to be areas of unrealised potential that could be developed to benefit construction stakeholders.

In addition, the following issues were also mentioned by the survey questionnaire respondents as having a very significant influence on the value gained by stakeholders from construction:

- Community consultation
- Choice of materials
- Quality of design
- Willingness for change
- Skill of project team
- Perception and expectations of the client

The following factors were also mentioned by the survey questionnaire respondents as enablers for construction stakeholders to gain appropriate value from construction:

- Location of project
- Project personnel: skills and culture
- Quality of the design
- Informing stakeholders of their gains

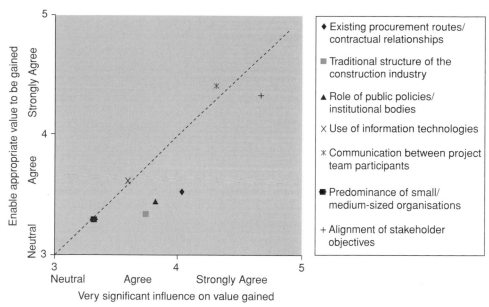

Figure A4.9   Factors influencing and enabling appropriate stakeholder value in construction.

2.4 Response to 'my organisation is generally *quick* to embrace new practices and procedures that could improve construction performance' and 'the construction industry in general is currently *radically improving* its performance'

The scatter graph (Figure A4.10) highlights that all the survey countries, with the exception of Canada, felt that at the organisational level they were quick to embrace new practices and procedures, but on the national front, the industry was not significantly improving its performance. The Canadians were the reverse of this: they felt that the industry was improving but were uncertain as to whether firms were quick to adopt these new techniques. The British and American responses closely reflect the mean response of all the survey countries.

The validity of the response that organisations are quick to respond but the industry is not changing can be questioned. Are the respondents biased towards the performance of their own organisation? These organisations, of course, could be leading their own research initiatives and/or are looking cross-sector and/or internationally for new improvement techniques. On closer examination of the results, the responses are dispersed across the selection criteria – the respective standard deviations are all seemingly high (Table A4.3). Therefore, it would appear that the respondents were being objective in their response, and so the results in Figure A4.10 can be relied upon.

It would seem that the response is more a reflection of a reactive industry where individual players do respond quickly, but the cumulative impact for the whole industry is not achieved. In short: the whole is less than the sum of the parts!

Table A4.3 Standard deviations (SD) of the results to improvement of organisation and industry performance.

Country	Organisational performance	Industry performance
Australia	0.983327	0.890871
Canada	0.866025	0.748331
Singapore	0.828459	1.052329
UK	1.187902	0.801811
USA	1.060275	0.900254

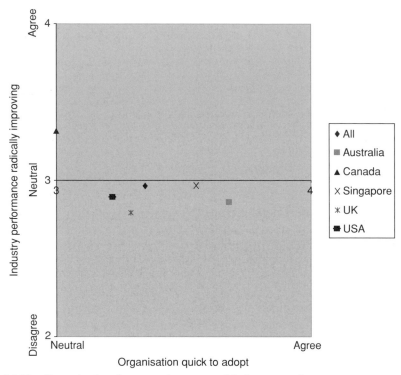

Figure A4.10 Organisational responsiveness and industry performance.

2.5 Response to '(the following) factors significantly inhibit the widespread adoption of new practices that could improve construction performance'

A number of issues were listed in the survey questionnaire as barriers that may impede the uptake of new construction practices. The responses are displayed in Figure A4.11. Only one of the issues listed was considered not to be an inhibitor – strict health and safety requirements. It is also interesting to note the ranking order of the issues questioned – these can be seen as priority areas for revaluing construction research. The actions of the client/industry and professional bodies protecting their

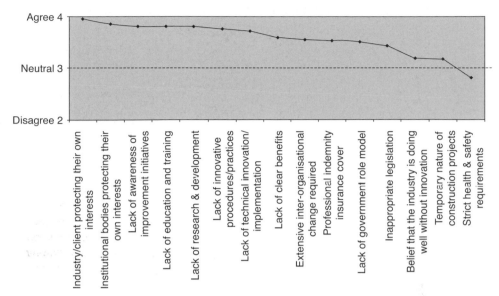

Figure A4.11   Factors that significantly inhibit the widespread adoption of new construction practices.

own interest were seen to be the greatest inhibitor, and at the other end of the scale is the temporary nature of construction projects. This citation has often been attributed to the lack of innovation by many government and institutional reports worldwide, but perhaps this has now been accepted as a characteristic of the industry. One respondent clearly stated that '... the biggest impairment to innovation is the restrictive building codes and the Government's slow initiation and acceptance of new materials and methods... also fear of litigation forces designers to avoid risk at all cost'. In addition, the following factors were also mentioned by the respondents:

- Low profit margins restricting the available resources for improvement
- Failure to address local conditions
- Industry/organisation does not reward innovators
- Poor and ineffective leadership
- Inappropriate culture

This result, however, somewhat contradicts the response to Section 2.3, where the respondents earlier indicated that the structure, procurement, use of IT, size of organisations in the industry, etc. enabled stakeholders to gain appropriate value from their involvement in construction. Figures A4.3 through to A4.8 suggest that construction stakeholders are satisfied with the nature of the industry as it stands today and, daresay, that there is no immediate need for new construction practices that enable them to gain value from construction. However, this stance does not seem to follow the general consensus of numerous governmental and institutional reports, discussion forums, national and international policy documents, the emergence of new tools and techniques – which all clearly advocate the need for change and improvement of the industry. Further, a number of respondents made additional comments to this effect:

- '...the construction industry is a blame culture driven by commercial and contractual "experts" and in it has driven the profit and performance margins down.'
- '...an emphasis/dominance on value for money have affected construction performance. Naturally, the lowest tender will be awarded...for the architects and engineers, the professional fee has been at a new low where they are not able to spend much time on the design before overshooting the fee.'
- A Canadian manufacturer/supplier reported: '...the integrated general contractor has long gone. General contractors only perform construction management or project management roles and pass all the risks to sub trades. Designers do not undertake specialist design functions, and again these are passed to sub trades. The whole risk/reward balance in construction has lost its meaning. The lower counterparts of the supply chain now carry all the risk.'
- '...the construction industry is too disintegrated in the delivery chain: too many people involved, lack of repeats in experiences, low wages, long hours of working, poor work place – it is a factory without a roof!'
- According to an Australian designer: '...union interference is the greatest threat to the construction industry.'
- A UK manufacturer/supplier stated that '...the construction industry has trade bodies which do not encapsulate all the needs of the client down to site end users. They do not engage with the lower echelons of the industry (subcontractors, SMEs, suppliers, etc.).'

Detailed analysis of the response to this section of the survey questionnaire by discipline would suggest that all the stakeholder groups were aligned to the general response (see Figure A4.12), particularly with the agreement that there was a general lack of innovative procedures/practices within the industry. The response to 'strict health and safety requirements' as an inhibiting factor to the widespread adoption of new initiatives was not unequivocally thought of as a barrier by the contractor, design team and client/end-user stakeholder groups. Perhaps unsurprisingly so, the manufacturers/suppliers felt that this was a barrier, as products/components have to conform to any new health and safety legislations – and new regulations frequently emerge and may restrict or stifle innovation. One contractor, in his additional comments, captured this stance: '...many times we feel that we are stretching or modifying the typical use of a material...usually, the supplier or maker of the product is not able to analyse or support our request...it is very hard to analyse an industry building everything from barns to high-rise buildings...it is hard to stay aware of product development worldwide...it is hard to find a producer that will discuss unusual uses for their product or slight modifications...it is hard to find contractors who will work with you on innovative methods or materials...it is very hard to find lenders to support unusual but valuable innovation.' Another contractor thus concluded: '...stakeholders have to work closer together to promote and use/test new developments.'

The response from the contractors and design team (who are positioned directly within the industry) were very similar across all of the criteria, whereas the

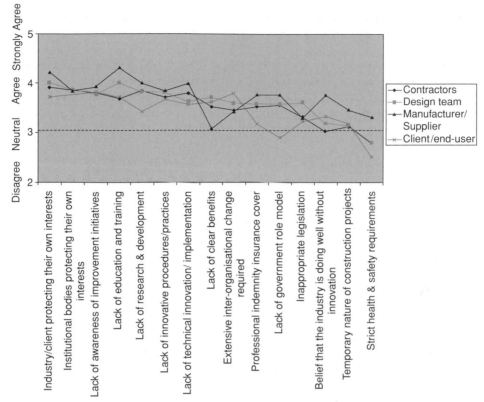

Figure A4.12 Factors of significance inhibiting the widespread adoption of new construction practices by discipline.

manufacturer/supplier and client/end-user response was more dispersed (these stakeholders are generally considered to be indirectly involved in the industry). Discipline-specific responses at the periphery of the general trend include the manufacturer/supplier response to 'lack of education and training' and 'lack of clear benefits', which may be attributed to the fact that they are sited lower down the construction supply-chain. In addition, the clients/end-users believe that a lack of a government role model should not hinder the adoption of new tools and techniques. Instead, perhaps, the organisations themselves should be driving innovation rather than waiting for government assistance – a global example is McDonalds who have developed their own product and process models and can build a new store within hours.

2.6 Response to '(the following are) effective drivers for *significantly increasing* the value delivered by construction'

Figure A4.13 lists a number of factors that could significantly increase the value delivered by construction – the respondents all agreed to the listing. Notably, learning from

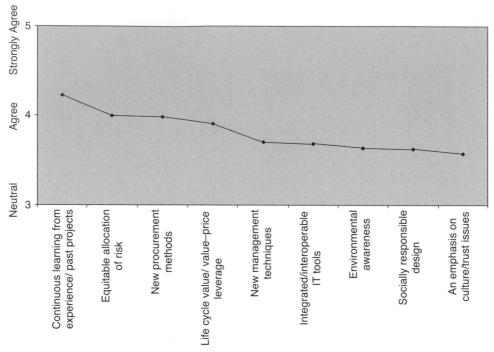

Figure A4.13 Drivers that could significantly increase the value delivered by construction.

past projects and the equal allocation of risk were seen to be areas for the concentration of research effort. In addition, the following factors were also mentioned by the respondents as significantly increasing the value of construction:

- Materials development
- Professional fees
- Legislation
- Improved briefing process
- Health and safety
- Integrated project team

Again, by analysing the findings by discipline (see Figure A4.14), the stakeholder groups generally concur in their response, with the minor exception of the manufacturer/supplier in some instances. Once more with the 'emphasis on culture/trust issues', and the 'equitable allocation of risk' criterion, the responses were more greatly spread – with the design team and the manufacturers/suppliers responses at the periphery of the scale. This spread may be attributed to the manufacturers/suppliers lower-down-the-supply-chain's role in the design and construction process, at the opposing end to the design team, who do not usually deal with softer issues such as culture and risk.

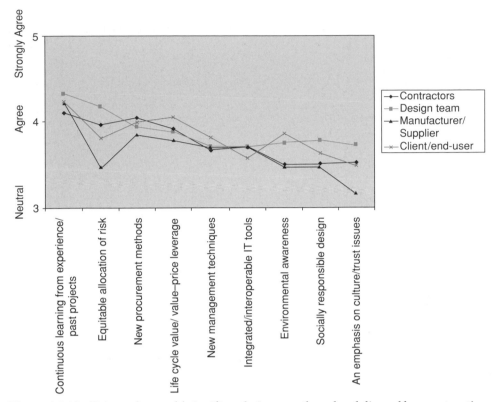

Figure A4.14 Drivers that could significantly increase the value delivered by construction by discipline.

2.7 Response to 'the construction industry is currently radically improving its performance (in the following areas)'

Out of the issues listed, the majority of survey questionnaire respondents agreed that the industry is radically improving its performance (see Figure A4.15). Some of the issues listed were earlier identified as existing barriers to construction improvement initiatives, and therefore it is positive to learn that these areas are being addressed. However, the topics of research and development, the industry's public image, and education and training still require a considerable amount of development application. A number of comments were also mentioned by the respondents to be possible areas of research:

- One respondent also pointed out that the soft issues of valuing construction practice must also be considered, such as the appropriateness of the leader, his/her personality/style in ensuring the appropriate culture is achieved.
- '... need to research on what value is for customers... value lies in the use of buildings, not merely in their existence... facilities management is a major part of the value stream.'
- '... not enough emphasis on design quality – resulting in inefficiencies.'

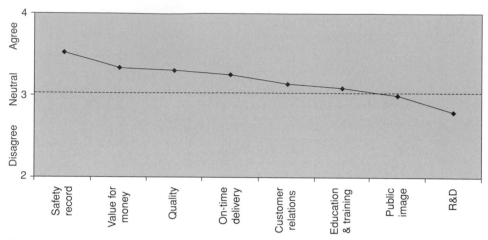

Figure A4.15   The radical improvement achieved in construction performance.

In addition, a number of concerns were also raised by the respondents to the status of the industry in the survey countries:

- A UK designer stated: '... there is a significant drift backwards to pre-2000 adversarial and lowest cost rather than lowest + value as the Egan/Latham/Construction Excellence message gets forgotten. Rethinking construction is only an awareness at senior management level, it hasn't cascaded down to office and site levels. Surveyors have far too much influence in the construction chain. Architects are largely blissfully unaware of the necessity to be team members. Local best practice clubs are poorly attended and sadly there is widespread disinterest in change improvement in the building industry.'
- A designer quoted: '... construction in Singapore is experiencing the worst downturn. The implementation of design and build contracts is changing the way architecture is being practised... the introduction of IT speeds everything and we have to respond accordingly, putting ourselves at risk on legal and procurement matters.'
- According to a Singaporean client/end-user: '... the construction work force depends largely on foreign import (80%) as many are transient workers or unskilled. Therefore, job site management has not improved much over 30 years. There were distinct differences between foreign and local contractors in terms of management, finance, innovation, R&D, communication, etc. 70% of construction methods are prefabricated in Australia, as compared to 20% in Singapore.'

2.8 Overall conclusions

From the responses of the international revaluing construction survey questionnaire, the following conclusions can be drawn:

- Perceptions of issues of revaluing construction are broadly similar across the survey countries.

Table A4.4 Revaluing construction research topics.

	Influencing & enabling value	Value inhibitors	Value drivers	Performance
Stakeholder objectives/benefits	X	X	X	
Procurement/contracts/risk allocation	X	X	X	
Role of public policies/ institutional bodies/ government	X	X		
Structure & image of the industry	X			X
Education & training		X		X
Research & development		X		X
Innovative procedures/practices (awareness & adoption examples)		X	X	
Culture/change issues		X	X	
Legislation		X		
Learning from past projects			X	
Environmental/social awareness			X	

- Perceptions of issues of revaluing construction are generally similar across the stakeholder groups positioned within the industry (contractors and the design team) and those who are indirectly involved (manufacturers/suppliers and clients/end-users).
- A concentration of effort is required to actively involve stakeholders who are usually considered to be indirectly involved in the construction industry, such as manufacturers/suppliers and clients/end-users.
- The key stakeholders in the construction industry are providing value to and gaining value from their sector. Indeed such is the nature and structure of the sector that it not only influences value but enables value to be gained.
- In Australia, Singapore, the UK and USA, organisations are generally embracing new practices and procedures, but nationally the industry is not significantly improving its performance. Canada was the reverse of this trend: the industry is improving but the respondents are somewhat uncertain as to whether their organisations are adopting these new initiatives. In short, it would seem that the response is more a reflection of a reactive industry where individual players do respond quickly, but the cumulative impact for the whole industry is not achieved. In short: the whole is *less* than the sum of the parts!
- The barriers to the adoption of innovative practices in the construction sector are perceived to be factors that can be found across all the survey countries, rather than the results of local circumstances, although there was a comment made on the nature of the Singaporean workforce. Hence, there is scope for international liaison such that is the purpose of this International Revaluing Construction study.
- The construction industry as a whole is generally improving its practice.

This international survey questionnaire sought to identify topics for a revaluing construction research agenda that the CIB might seek to fill; these issues are summarised in Table A4.4. The gaps identified provide a positive foundation upon which the follow-up international workshops will build in order to develop holistic future CIB actions in this area.

References

Abbott, P.G. (1985) *Technology Transfer in the Construction Industry*. London, The Economist Intelligence Unit.

Abdel-Razek, R. & McCaffer, R. (1987) A change in the UK construction industry structure: implications for estimating. *Construction Management & Economics* **5**: 227–242.

Agryris, C. & Schon, D. (1978) *Organisational Learning: A Theory of Action Perspective*. Reading, Mass., Addison-Wesley.

Ahmad, R. (2004) KL fines contractor $35m for project delay. *The Straits Times* (Singapore) **October 23**: 21.

Alarcón, L. (1997) *Lean Construction*. The Netherlands, A.A. Balkema, p. 497.

Anderson, E. & Jap, S.D. (2005) The dark side of close relationships. *MIT Sloan Management Review* **46**(3): 75–82.

Ang, G.K.I., Prins, M., & de Ridder, H.A.J. (2003) *The Added Value of Architecture in Performance-Based Control of Design & Construct Contracts* – Singapore, CIB Symposium on Knowledge Construction.

Ang, G.K.I. (2004a) *Meeting Report 11th International Client Platform*, 23–24 September 2004, The Hague. CIB website: www.cibworld.nl; www.PSIBouw.nl

Ang, G.K.I. (2004b) *Competing Revaluing Construction Paradigms in Practice*, CIB Report. The Netherlands, CIB.

Ang, G.K.I., Wyatt, D.P. & Hermans, M.H. (2001) *A Systematic Approach to Define Client Expectations of Total Building Performance during the Pre-Design Stage*. *Proceedings of CIB World Congress, Wellington, New Zealand, 2001*, Stream "Client Expectations of Performance, and Methods of its Measurement".

Ang, G.K.I., Courtney, R., Spekkink, D., et al. (2004) *Inventory of International Reforms in Building and Construction* (PSIBouw PP1). The Netherlands, CUR Gouda, June 2004. Website: www.PSIBouw.nl

Ang, Y.K. & Ofori, G. (2001) Chinese culture and successful implementation of partnering in Singapore's construction industry. *Construction Management and Economics,* **19**: 619–632.

Ansoff, H. (1970) *Corporate Strategy*. London, Pelican.

Applebaum, H.A. (1981) Royal blue: The culture of construction workers. In: *Case Studies in Cultural Anthropology* (ed. G. Spindler & L. Spindler). Stanford University, Holt, Rinehart and Winston International.

Arbulu, R.J., Tommelein, I.D., Walsh, K.D. & Hershauer, J.C. (2003) Value stream analysis of a re-engineered construction supply chain. *Building Research & Information* **31**(2): 161–171.

Artto, K.A., Kahkonen, K. & Pitkanen, P.J. (2000) *Unknown Soldier Revisited: A Story of Risk Management*. Helsinki, PMA.

Audretsch, D.B., Baumol, W.J. & Burke, A.E. (2001) Competition policy and dynamic markets. *International Journal of Industrial Organisation* **19**: 613–634.

Australian Procurement and Construction Council (1997) *Construct Australia.* Canberra, APCC.

AWT (1997) *De invloed van wet — en regelgeving op innovatie.* The Hague, Advisory Council for Technology and Science Policy, AWT publication no. 27.

Bain, J.S. (1956) A note on pricing in monopoly and oligopoly. *American Economics Review* **39**: 448–469.

Bakens, W., Foliente, G. & Jasuja, M. (2005) Engaging stakeholders in performance-based building: lessons from the Performance-Based Building (PeBBu) Network. *Building Research & Information* **33**(2): 149–158.

Bakis, N., Aouad, G. & Kagioglou, M. (2005) Computer integrated design technology in the construction industry: Progress so far and future challenges. In: *Proceedings of the International Built and Human Environment Research Week 2005,* Salford.

Ball, M. (1988) *Rebuilding Construction: Economic Change in the British Construction Industry.* London, Routledge.

Ballantine, J. & Stray, S. (1999) Information systems and other capital investments: Evaluation practice compared. *Logistics and Information Management* **12**(1–2): 78–93.

Ballantine, J., Galliers, R. & Stray, S. (1994) Information systems/technology investment decisions: The use of capital investment appraisal techniques in organisations. *Proceedings of the First European Conference on Information Technology Investment Evaluation* (ed. A. Brown & D. Remenyi), 13–14 September 1994, Henley on Thames, Oxon.

Ballard, G. & Howell, G.A. (2003) Lean project management. *Building Research & Information* **31**(2): 119–133.

Bannister, F. & Remenyi, D. (1999) Value perception in IT investment decisions. *The Electronic Journal of Information Systems Evolution* **3**: Paper 1.

Banwell, H. (1964) *Report of the Committee on the Placing and Management of Contracts for Building and Civil Engineering Work.* London, HMSO.

Barlow, J., Childerhouse, P., Gann, D., Hong-Minh, S., Naim, M. & Ozaki, R. (2003) Choice and delivery in house building: Lessons from Japan for UK house builders. *Building Research & Information* **31**(2): 134–145.

Barrett, P.S. (2001) A survey of construction clients' needs. Paper presented at the *CIB World Building Congress,* 2–6 April, Wellington, New Zealand.

Barrett, P. (2002) Innovation in construction – what needs to change? In: *Proceedings of the CIB W55/W65 Joint 10th International Symposium on Construction Innovation and Global Competitiveness,* 9–13 September, University of Cincinnati.

Barrett, P.S. (2003) Construction management pull for ND CAD. In: *4DCAD and Visualization in Construction.* The Netherlands, Swets & Zeitlinger., pp. 253–270.

Barrett, P.S. & Barrett, L.C. (2003) Research as a kaleidoscope on practice. *Construction Management and Economics* **21**(7): 755–766.

Barrett, P.S. & Barrett, L.C. (2004) *Revaluing Construction: Final Synthesis Report on International Workshops.* The Netherlands, CIB.

Barrett, P.S. & Gilkinson, N.R. (2004) The unanticipated impacts of research on practice. *Proceedings of the CIB World Building Congress,* 2–7 May, Toronto, Ontario, Canada. The Netherlands, CIB.

Barrett, P.S. & Lee, A. (2004) *Revaluing Construction: International Survey Questionnaire Results.* The Netherlands, CIB.

Barrett, P.S. & Sexton, M. (1999) *The transformation of 'out-of-industry' knowledge into construction industry wisdom.* London, CRISP Motivation Theme Group.

Barrett, P.S. & Stanley, C. (1999) *Better Construction Briefing*. Oxford, Blackwell Science.

Barrett, P.S., Sun, M., Petley, G.J., Ali, K.N. & Kagioglou, M. (2003) *More Productive Minor Construction Projects Through Information Technology (MoPMIT)*. Salford, Salford Centre for Research & Innovation in the Built & Human Environment.

Barry, F. (1997) Convergence is not automatic: Lessons from Ireland for Central and Eastern Europe. *Economic Journal, Ireland*.

Beer, S. (1985) *Diagnosing the System for Organisations*. Chichester, John Wiley.

Bennett, J. & Jayes, S. (1998) *The Seven Pillars of Partnering: A Guide to Second Generation Partnering*. London, Thomas Telford.

Betts, M. & Ofori, G. (1993) Strategies for technology push: Lessons for construction innovation. *Construction Management and Engineering*, ACSE, 454–456.

Bilbo, D., Fetters, T., Burt, R. & Avant, J. (2000) A study of the supply and demand for construction education graduates. *Journal of Construction Education* 5(1): 78–89.

Blyth, A. & Worthington, J. (2001) *Managing the Brief for Better Design*. London, Spon Press.

Bon, R. (1990) *The World Building Market 1970-85*, Building Economics and Construction Management: Proceedings of the CIB W65 Symposium, Sydney.

Bon, R. (1992) The future of international construction: Secular patterns of growth and decline. *Habitat International* 16(3): 119–128.

Bower, D. (2003) The role of procurement in the construction industry. In: *Management of Procurement* (ed. D. Bower). London, Thomas Telford.

Briscoe, G., Dainty, A.R.J. & Millet, S.J. (2000) The impact of the tax system on self-employment in the British construction industry. *International Journal of Manpower* 21(8): 596–613.

Brouseau, E. & Rallet, A. (1995) Efficacité et Inefficacité de L'organisation du Bâtiment: Une Interprétation en Termes de Trajectoire Organisationnelle. *Revue d'Economie Industrielle* 74(4): 9–30.

Broyd, T. (2001) *Constructing the Future*. London, CRISP.

Brynjolfsson, E. (1993) The productivity paradox of information technology. *Communications of the ACM,* **36**: 12: 66–77.

Building Futures (2001) *2020 Vision: Our Future Healthcare Environments, Joint CABE/RIBA Report*. www.buildingfutures.org.uk

Building Futures Council (2000) *The Future of the Design and Construction Industry (Projection to 2015)*. Washington D.C., CERF.

Building Policy Task Force (2000), *The Danish Construction Sector in the Future—from Tradition to Innovation*. The Netherlands, Ministry of Housing and Urban Affairs, Ministry of Trade and Industry.

Burns, T. & Stalker, G.M. (1961) *The Management of Innovation*. London, Tavistock.

CABE Space (2004). *Is the Grass Greener? Learning from International Innovations in Urban Green Space Management*. London, CABE.

CABE (ed.) (2005) *Measuring the Impact of Architecture and Design on the Performance of Higher Education in Institutions*. London, Commission for Architecture and the Built Environment.

Cairncross, F. (2001) *The Death of Distance. 2.0 How the Communications Revolution Will Change Our Lives*. London, Texere Publishing.

Cannon, J. (1994) Lies and construction statistics. *Construction Management and Economics* 12(4): 307–313.

Carassus, J. (ed.) (2004) *The Construction Sector System Approach: An International Frame-work*.Report by CIB W055-W065 Construction Industry Comparative Analysis Project Group. Rotterdam, CIB.

Carillo, P. (1994) Technology transfer: A survey of international construction companies. *Construction Management and Economics* **12**: 45–51.

Carrillo, P.M., Robinson, H.S., Anumba, C.J. & Ahmed, A.G.M. (2003) IMPaKT: a framework for linking knowledge management to business performance. *Electronic Journal of Knowledge Management* **1**(1): 1–12.

Carty, G. (1995) Construction. *Journal of Construction Engineering and Management* **121**(3): 319–328.

CCIC (2004) *The Canadian Construction Innovation Council*. Canadian Construction Innovation Council (CCIC), pp. 1–4.

CERF (2000) Guidelines for moving innovations into practice. Working Draft Guidelines for the CERF International Symposium and Innovative Technology Trade Show, Washington.

Chang, T. & Chien-Chung, N. (2004) A note on testing the causal link between construction activity and economic growth in Taiwan. *Journal of Asian Economics* **15**: 591–598.

Checkland, P. & Scholes, J. (1990) *Soft Systems Methodology in Action*. Chichester, John Wiley.

CIB (1999) *Agenda 21 on Sustainable Construction*. The Netherlands, CIB, CIB Report Publication 237.

CIB (2002) *Agenda 21 for Sustainable Construction in Developing Countries: A Discussion Document* (ed. CSIR). Pretoria, CIB.

CIB (2004) *The Construction Sector System Approach: An International Framework* (ed. J. Carassus). Rotterdam, International Council for Research and Innovation in Building and Construction.

CIBSE (1999) Technical memorandum TM22: 1999, Energy assessment and reporting methodology: office assessment method. London, Chartered Institution of Building Services Engineers.

CIDA (1995) *Measuring Up or Muddling Through: Best Practice in the Australian Non-Residential Construction Industry*. Sydney, Australia, Construction Industry Development Agency and Master Builders Australia.

CII (1999) *Vision 2020*. Texas, CII.

CII (ed.) (2004) *Re-engineering the Construction Delivery Process*. Queensland, Queensland University.

Cloutier, L. (2004) La construction veut contrer un retard de 10 ans au chapitre de l'innovation. In *La Presse Affaires*, Montreal, pp. 24–25.

Construct IT(ed.) (1998a) Benchmarking Best Practice Reports: Supplier Management and Project Programming and Control. Salford, University of Salford.

Construct IT (1998b) *Measuring the Benefits of IT Innovation*. Construct IT Centre of Excellence, ISBN 1-900491-7.

Construction Industry Review Committee (2001) *Construct for Excellence*. Hong Kong.

Construction 21 Steering Committee (1999) *Re-Inventing Construction*. Singapore, Ministry of Manpower and Ministry of National Development.

Consultants, E.R. (2004) Effective teamwork. In: *A Best Practice Guide for the Construction Industry*. Watford, CBP, p. 16.

Courtney, R. & Winch, G.M. (2003) Re-engineering construction: the role of research and implementation. *Building Research & Information* **31**(2): 172–178.

Courtney, R., Spekkink, D., Ang, K.I. (2004) *Inventory of international reforms in building and construction* June 2004, PSI Bouw website: http://www.psibouw.nl/upload/documents/ Engelse%20Projecten/Inventory%20of%20international%20reforms%20in%20building %20and%20construction.pdf

CPN (ed.) (2004) Value Management in Construction: Examples from the UK and Mainland Europe. CPN: Construction Productivity Network, pp. 1–10.

Cyon Research (2003) *Building Information Model: A Look at Graphisoft's Virtual Building Concept*. Bethesda, MD, Cyon Research Corporation.

Dainty, A.R.J., Bryman, A. & Price, A.D.F. (2002) Empowerment within the UK Construction Sector. *Leadership & Organisation* **23**(6): 333–342.

Davey, C.L., Powell, J.A., Cooper, I. & Powell, J.E. (2004) Innovation, construction SMEs and action learning. *Engineering, Construction and Architectural Management* **11**(4): 230–237.

Davidson, C. & Dimitrijevic, B. (2004) *Agenda 21: Information and Documentation – a Research Agenda*. A report for the CIB Programme Committee. CIB.

Dawood, N., Sriprasert, E., Mallasi, Z. & Hobbs, B. (2002) Development of an integrated information resource base for 4D/VR construction processes simulation. *Automation in Construction* **12**: 123–131.

Department of Economic and Social Affairs (1962) Report of the Ad-Hoc Group of Experts on Housing and Urban Development. New York.

de Ridder, H.A.J. (2001) Dynamic performance control concept of design & construct of complex systems. *Volume Two Proceedings of CIB World Building Congress 2001, Performance in Product and Practice*, 2–6 April 2001, Wellington, New Zealand, pp. 299–308.

de Ridder, H.A.J. (2002) *Performance Based Control of Design & Construct Contracts*, International Conference, 6-8 May 2002, Hong Kong, CIB Report, Publication 283, pp. 75–86.

Dewulf, G. & van Meel, J. (2004) Sense and nonsense of measuring design quality. *Building Research & Information* **32**(3): 247–250.

DIST (1999) *Building for Growth: A Draft Strategy for the Building and Construction Industry*. Canberra, Australia, Department of Industry, Science and Tourism, Commonwealth of Australia Publication.

Doree, A.G., Holmen, E.K.H.S., & Caerteling, J. (2003) Competitive relationships in construction – competing with the same counterpart. *Proceedings of the Annual Conference of the Association of Researchers in Construction Management*, 3–5 September 2003, Brighton, pp. 817–826.

Doz, Y.L. (1996) The evolution of cooperation in strategic alliances: initial conditions or learning processes? *Strategic Management Journal* **17**: 55–83.

Drewer, S. (2001) A perspective on the international construction system. *Habitat International* **25**(1): 69–79.

DTI (2003a) *Construction Statistics Annual* (2003 edn). London, HMSO.

DTI (2003b) *UK 2001 Census Statistics*. London, HMSO.

DTI (2006) *Output and Employment in the Construction Sector: First Quarter 2006*. www.dti.gov.uk/sectors/construction/ConstructionStatistics/OutputandEmployment/ quarter106/page29702.html (downloaded 26 March 2007).

Dubois, A. & Gadde, L.E. (2002) The construction industry as a loosely coupled system: implications for productivity and innovation. *Construction Management and Economics* **20**: 621–631.

Duffy, F. with Hutton, L. (1998) *Architectural Knowledge: The Idea of a Profession*, London, E+FN Spon.

Duncan, J. (2005) Performance-based building: lessons from implementation in New Zealand. *Building Research & Information* **332**: 120–127.

Ebohon, J.O. (2003) The knowledge-based global economy: Implications for construction industry development in Sub-Saharan Africa. In: *Proceedings of the Joint International Symposium of CIB W55, W65 and W107 International Conference* (ed. G. Ofori & Y.Y.F. Ling), 22–24 October, Singapore, Vol. 2, pp. 521–533.

Eclipse Research Consultants (2002) *Better Designed Buildings: Improving the Valuation of Intangibles – A Literature Review*. Cambridge, Eclipse Consulting, pp. 1–17.

Economic Commission for Africa (1965) *Housing in Africa*. New York, United Nations.

Edkins, A. (2000) *Building Future Scenarios*. London, CRISP.

Edmonds, G.A. and Miles, D.W.J. (1984) *Foundations for Change: Aspects of the Construction Industry in Developing Countries*. London, Intermediate Technology.

Egan, J. (1998) Rethinking construction. Report from the Construction Task Force, Department of the Environment, Transport and Regions. London.

Eley, J. (2004) Forum: Design quality in buildings. *Building Research & Information* **32**(3): 255–260.

Emmerson, H. (1962) *Studies of Problems before the Construction Industries*. London, HMSO.

Euroconstruct (2004) *Prospects for European Markets*. Paris, Euroconstruct.

European Commission (2003) *2003 Observatory of European SMEs*. Luxembourg, European Commission.

Fairclough, J. (2002) *Rethinking Construction Innovation and Research: A Review of Government R&D Policies and Practices*. Department of Transport Local Government Regions. London, HMSO.

Farbey, B., Land, F. & Targett, D. (1993) Information technology – why evaluate? In: *IT Investment: A Study of Methods and Practice*. Oxford, Butterworth-Heinemann, pp. 12–22.

Farmer, D. (1997) Purchasing myopia revisited. *European Journal of Purchasing and Supply Management* **3**(1): 1–8.

Fellows, R., Langford, D., Newcombe, R. & Urry, S. (1983) *Construction Management in Practice*. New York, Longman.

FIEC (2003) Construction in Europe: Key Figures. Leaflet, August.

Fischer, G.-N. & Vischer, J.C. (1998) *L'évaluation des environnements de travail, la méthode diagnostique Montréal*. Presses de l'Université de Montréal and Brussels, DuBoek.

Flanagan, R., Jewell, C., Larsson, B. & Sfeir, C. (2000) *Vision 2020*. Sweden, Chalmers University.

Fleming, A., Larsen, G. & Kao, C. (2006) The development of a competitive index for construction firms to enable informed procurement. Paper submitted to the *CIB W92 Conference*, 29 Nov.–1 Dec. 2006, University of Salford.

Flyvbjerg, B., Bruzelius, N. & Rothengater, W. (2003) *Mega Projects and Risk: An Anatomy of Ambition*. Cambridge, Cambridge University Press.

Fong, P.S. (1999) Organisational knowledge and responses of public sector clients towards value management. *The International Journal of Public Sector Management* **12**(5): 445–454.

Fortune, C. & Lees, M. (1993) Early advice to clients – ignorance in practice, *Proceedings of the Annual Conference of ARCOM*, September 1993, Exeter College, Oxford University.

French, W.L. & Bell, C.H. Jr (1984) *Organisational Development: Behavourial Science Interventions for Organisation Improvement* (3rd edn). New Jersey, Prentice-Hall International Editions.

Future Purchasing Alliance (2003) *Connecting Purchasing and Supplier Strategies to Shareholder Value*. FPA, UK.

Galbraith, J.R. (1977) *Organisation Design*. Reading, Addison-Wesley.

Gale, A.W. & Fellows, R.F. (1990) Challenge and innovation: The challenge to the construction industry. *Construction Management & Economics*, **8**(4).

Gallaher, M.P., O'Connor, A.C., Dettbarn, J.L. & Gilday, L.T. (2004) *Cost Analysis of Inadequate Interoperability in the U.S. Capital Facilities Industry*. National Institute of Standards and Technology.

Gallicon (2001) *Supporting Partners: Industrial Report*. Salford, University of Salford.

Gann, D. (2003) Nanotechnology and implications for products and processes. London, CRISP.

Gann, D.M. (2000) *Building Innovation: Complex Constructs in a Changing World*. London, Thomas Telford.

Gibb, A. G. & Isack, F. (2003) Re-engineering through pre-assembly: client expect drivers. *Build Research and Information*. **31** (2): 146–160.

Gibbons, M., Limoges, C., Nowotny, H., Schwartzman, S., Scott, P. & Trow, M. (1994) *The New Production of Knowledge*. London, Sage.

Gibson, E.J. (1982) *Working with the Performance Approach in Building*. Rotterdam, CIB.

Gladwell, M. (2000) *The Tipping Point*. London, Abacus.

Graphisoft (2003) *The Graphisoft Virtual Building: Bridging the Building Information Model from Concept into Reality*. Graphisoft Whitepaper.

Gratton, L. & Ghoshal, S. (2005) Beyond Best Practice. *MIT Sloan Management Review* **46**(3): 49–57.

Gray, C. & Hughes, W. (2001) *Building Design Management*. Oxford, Butterworth-Heinemann.

Green, S.D. & May, S.C. (2003) Re-engineering construction: going against the grain. *Building Research & Information* **31**(2): 97–106.

Gronroos, C. (1994) From marketing mix to relationship marketing. *Management Decision* **32**(2): 1–22.

Gyles, R. (1992) *Royal Commission into Productivity in the Building Industry in New South Wales*. Sydney, Australia, New South Wales Government, Vols 1–10.

Halman, J.I.M. & Prins, M. (1997) *Virtual Reality in Architectural Design Management*. Eindhoven, The Netherlands, Eindhoven University of Technology.

Hampson, K. & Brandon, P. (2004) *Construction 2020: A vision for Australia's Property and Construction Industry*. Brisbane, Australia, CRC Construction Innovation.

Hansen, M.T., Nohria, N. & Tierney, T. (1999) What's Your Strategy for Managing Knowledge. *Harvard Business Review* **77**(2), pp. 106–116.

Hart, S. & Milstein, M.B. (2003) Creating sustainable value. *Academy of Management Executive* **17**(2): 56–69.

Harty, C.F., Goodier, C.I., Soetanto, R., Austin, S.A., Dainty, A.R.J., Price, A.D.F. (2006a) The futures of construction: a critical review of construction futures studies. *Construction Management and Economics* (in press).

Harty, C.F., Goodier, C.I., Soetanto, R., Austin, S.A., Dainty, A.R.J., Price, A.D.F. & Thorpe, A. (2006b) Scenario development: a methodology for aligning contemporary practices with the potential futures of UK construction. In: *Proceedings of ARCOM Twenty-second Annual Conference* (ed. D. Boyd), 4–6 September, Birmingham, pp. 601–610.

Heerwagen, J.H., Kampschroer, K., Powell, K.M. & Loftness, V. (2004) Collaborative knowledge work environments. *Building Research & Information* **32**(6): 510–528.

Hendriks, F. (2004) The poison is the dose or how 'more egalitarianism' may work in some places but not in all. *European Journal of Social Science Research* **17**(4): 349–361.

HESA (Higher Education Statistics Agency) (2006) *Student Tables.* www.hesa.ac.uk/holisdocs/pubinfo/stud.htm (accessed 26 March 2007).

Hillebrandt, P.M. (1999) Problems of larger local contractors: Causes and possible remedies. Paper presented at the *Second Meeting of CIB TG29,* 25–26 June, Kampala, Uganda.

Hillebrandt, P.M. (2000) *Economic Theory and the Construction Industry* (3rd edn). London, Macmillan.

Hills, G. & Tedford, D. (2003) The education of engineers: the uneasy relationship between engineering, science and technology. *Global Journal of Engineering Education* **7**(1).

HKSA (Hong Kong Ethics Development Centre and Hong Kong Society of Accountants) (1997) *Ethics in Management: A Practical Guide for Professional Accountants.* Hong Kong.

HMSO (1964) *The Placing and Management of Contracts for Building and Civil Engineering Works (The Banwell Report).* London, HMSO.

HMSO (1994) *Constructing the Team: Joint Review of Procurement and Contractual Arrangements in the United Kingdom Construction Industry.* Final Report. London, HMSO.

Horgen, T.H., Joroff, M.L., Porter, W.L. & Schon, D.A. (1999) *Excellence by Design Transforming Workplace and Work Practice.* New York, John Wiley.

Houvinen, P. (2004) Applied business-management research: How do we incorporate this missing link into our revaluing construction agenda? Presented at the *CIB World Building Congress – Building for the Future,* 2–7 May, Toronto, Ontario, Canada.

Howell, D. (1999) Builders get the manufacturers in. *Professional Engineer* May, 24–25.

HSE (1998) How big firms manage contractual relations to reduce risk. In: *Managing Risk – Adding Value.* London, Health & Safety Executive, p. 82.

Hughes, W. (2003) De-professionalized, automated construction procurement. In: The Professionals' Choice: The Future of the Built Environment Professions. Build Futures report for CABE/RIBA.

ICCF International Client Platform (2004) Meeting Report, 23–24 September 2004, www.iccf.nl

IAF (Institute for Alternative Futures) (2005) *The 2029 Project: Achieving an Ethical Future in Biomedical R&D.* Alexandria, Virginia, USA, www.altfutures.com

International Labour Office (1987) *Guidelines for the Development of Small-scale Construction Enterprises.* Geneva, ILO.

International Labour Organization. (2004) *Global Distribution of Construction Output and Employment,* www.ilo.org/public/english/dialogue/sector/sectors/constr/global.htm (accessed Nov 2004).

Jaunzens, D., Warriner, D., Garner, U. & Waterman, A. (2001) *Applying facilities expertise in building design.* London, DTI.

Johnson, P. & Duberley, J. (2000) *Understanding Management Research.* London, Sage.

Johnston, R, (2004) Towards a better understanding of service excellence. *Managing Service Quality* **14**(2/3): 129–133.

Jonsen, A.R. & Toulman, S. (1988). *The Abuse of Casuistry.* Berkeley, CA, University of California Press.

Kagioglou, M., Cooper, R., Aouad, G., Hinks, J., Sexton, M. & Sheath, D. (1998) *Final Report: Generic Design and Construction Process Protocol.* The University of Salford.

Kaplan, R.S., & Norton, D.P. (1992) The balance scorecard:measure and drive performance, *Harvard Business Review,* Jan-Feb, 71–80.

Kashiwagi, D. & Verdini, W. (2004) *Best Value Procurement, Outsourcing, Supply Chain Management, and Six Sigma Application to the Delivering of Construction.* Arizona State University.

Kast, F.E. (1985) *Organisation and Management: A Systems Contingency Approach.* New York, McGraw Hill.

Kast, F.E. & Rosenzweig, J.E. (1973) *Contingency Views of Organization and Management.* Palo Alto, Science Research Associates.

Kay, J. (1993) *Foundations of Corporate Success.* Oxford, Oxford University Press.

Kelly, J. (2001) Goodbye Cardinal Newman, hello David Beckham: The rise & rise of vocationalism in university curricula, John Kelly, *Industry and Higher Education*, December 2001.

Kelly, J. & Hunter, K. (2003) *Best Value: the Three Wheels.* Glasgow, RICS Education Trust, School of Built and Natural Environment, Glasgow Caledonian University.

Keuning, S. (2000) Some thoughts on the impact of globalisation on statistics. In: Official Statistics in the 20th Century: Landmarks and Challenges, *Proceedings of the 85th Directors General of National Statistics Conference*, Brussels.

Kieran, S. & Timberlake, J. (2004) *Refabricating Architecture.* New York, McGraw-Hill.

King, Martin L. (1948) *The Purpose of Education.* Morehouse College, www.toptags.com/aama/voices/speeches/pofed.htm (accessed Nov. 2004).

Kirmani, S.S. (1988) The Construction Industry in Development: Issues and Options. Discussion Paper, Infrastructure and Urban Development Department. Washington D.C., World Bank.

Kirmani, S.S. & Baum, W.C. (1992) *The Consulting Professions in Developing Countries: a Strategy for Development.* Washington D.C., World Bank.

Kirzner, I.M. (1973) *Competition and Entrepreneurship,* Illinois, University of Chicago Press.

Koskela, L. (2003) Is structural change the primary solution to the problems of construction? *Building Research & Information* 31(2): 85–96.

Kotter, J. (1996) *Leading Change.* Cambridge, Mass, Harvard Business School Press.

Kunz, J., Fischer, M., Haymaker, J. & Levitt, R. (2002) Integrated and automated project processes in civil engineering: experiences of the Centre for Integrated Facility Engineering at Stanford University, *Computing in Civil Engineering Proceedings*, ASCE, Reston, VA, 96–105, January 2002.

Lahdenpera, P. (1995) *Reorganizing the building process: The holistic approach.* Espoo, Juilkaisija-Utgivare.

Lambert, R. (2003) *Lambert Review of Business – University Collaboration.* London, HM Treasury (available on line at www.lambertreview.org.uk)

Langdon, D. & Seah International (2004) *World Construction Review/Outlook 2003/4.* www.davislangdon.com (accessed 16 December 2004).

Langford, D.A. & Male, S. (1992) *Strategic Management in Construction.* Aldershot, Gower.

Lansley, P., McCreadie, C., Tinker, A., Flanagan, S., Goodacre, K. & Turner-Smith, A. (2004) Adapting the homes of older people: a case study of costs and savings. *Building Research & Information* 32(6): 468–483.

Lansley, P.R. (1987) *Managerial Skill and Corporate Performance in the Construction Industry. Managing Construction Worldwide*: Volume II: Productivity and Human Factors in Construction. Ascot, Berks, Chartered Institute of Building Englemere.

Larsen, J.N. (2001) Knowledge, human resources and social practice: the knowledge-intensive business service firm as a distributed knowledge system. *The Services Industries Journal* **21**(1): 81–102.

Latham, M. (1994) *Constructing the Team: Joint Review of Procurement and Contractual Arrangements in the UK Construction Industry.* Department of the Environment, HMSO.

Lawrence, P.R. & Lorsch, J.W. (1967) *Organization and Environment: Managing Differentiation and Integration.* Boston, Harvard University Press.

Lawson, B. (1990) *How Designers Think* (2nd edn). Oxford, Butterworth Architecture.

Leaman, A. & Bordass,W. (2004). Phase 5, occupancy and post-occupancy evaluation. In: *Building Performance Assessment* (ed. W.F.E. Preiser & J.C. Vischer). Oxford, Elsevier Science Publishers.

Lee, A. & Barrett, P.S. (2006) Value in construction: An international study. *International Journal of Construction Management* **6**: 81–95.

Lee, A., Kagioglou, M., Cooper, R. & Aouad, G. (2000) Production management: The process protocol approach. *Journal of Construction Procurement* **6**(2): 164–183.

Lee, A., Marshall-Ponting, A.J., Aouad, G., Wu, S., Koh, I., Fu, C., Cooper, R., Betts, M., Kagioglou, M. & Fischer, M. (2003) *Developing a Vision of nD-Enabled Construction. Construct IT Report.* Salford, University of Salford.

Lee, A., Wu, S., Marshall-Ponting, A., Aouad, G., Cooper, R., Tah, J.H.M., Abbott, C. & Barrett, P.S. (2005) *nD modelling Road Map, a Vision for nD Enabled Construction.* Centre for Construction Innovation-Salford.

Lees, M. (2005) *Long-term Educational and Technological Implications of Revaluing Construction,* CIB Report. The Netherlands, CIB.

Liu, A.M.M. & Fellows, R. (1999) Cultural issues. In: *Procurement Systems: A Guide to Best Practice in Construction* (ed. S. Rowlinson & P. McDermott). London, Spon, pp. 141–162.

Liu, A.M., Fellows, R. & Ng, J. (2004) Surveyors' perspectives on ethics in organisational culture. *Engineering, Construction and Architectural Management* **11**(6): 438–449.

Love, P.E.D, Haynes, N.S. & Irani, Z. (2001) Construction manager's expectations and observations of graduates, *Journal of Managerial Psychology* 16(8): 579–593.

Luck, R. (2002) Dialogue in participatory design. Presented at the *Common Ground Conference,* September, University of Brunel.

Luna, M. & Velasco, J.L. (2003) Bridging the gap between firms and academic institutions: The role of 'translators'. *Industry and Higher Education* **October**.

Male, S. (2003) Future trends in construction procurement: procuring and managing demand and supply chains in construction. In: *Management of Procurement* (ed. D. Bower). London, Thomas Telford.

Manseau, A. (2003) *Survey of Major Clients: Improvements and Innovation in Construction Investments.* London, NRC.

Marmot, A., Eley, J., & Bradley, S. (2004) Phase 2: Programming/Briefing and Programme Review in Preiser, W. & Vischer, J. C. (eds) *Assessing Building Performance* Oxford, Elsevier Science Publishing.

Matlay, H. (2000) Industry – higher education collaborations within small business clusters, *Industry and Higher Education* **December**.

McDermott, P. (1999) Strategic and emergent issues in construction procurement. In: *Procurement Systems: A Guide to Best Practice in Construction.* London, E&FN Spon.

McDermott, P. (2006) Think piece: Policy through procurement? In: *The Future of Procurement and its Impact on Construction,* a Workshop of Joint Contracts Tribunal & the University of Salford, 19 July 2006.

McGregor, D.M. (1957) The human side of enterprise. In: *Management and Motivation* (ed. V.V.H.D.E. L.). Middlesex, Penguin Books Ltd, pp. 306–319.

McKeen, J.D., Smith, A.H. & Parent, M. (1999) An integrative research approach to assess the business value of information technology. In: *Measuring Information Technology Investment Payoff: Contemporary Approaches* (ed. M.A. Mahmood & E.J. Szewczak). Hershey, PA, Idea Group Publishing, pp. 5–23.

Mende, W.M., Brecht, L. & Osterle, H. (1994) Evaluating existing information systems from a business process perspective. *Proceedings of the ACM 1994 Computer Personnel Research Conference on Reinventing IS: Managing Information Technology in Changing Organizations*, April, Alexandria, VA, USA.

Miles, D. & Neale, R. (1991) *Building for Tomorrow: International Experience in Construction Industry Development*. Geneva, International Labour Office.

Miles, R.E. & Snow, Ch.C., 1986–1997, *Network Organisations: New Concepts for New Forms*, The McKinsey Quarterly, pp. 53–66.

Milford, R. (2004) Re-valuing sustainable construction. Presented at the *CIB World Building Congress – Building for the Future*, 2–7 May, Toronto, Ontario, Canada.

Ministry of Works of Tanzania (1977) *Local Construction Industry Study*. Dar es Salaam.

Moavenzadeh, F. & Hagopian, F. (1984) The construction industry and economic growth. *Asian National Development* **June/July**: 56–60.

Mohsini, R.A. & Davidson, C.H. (1992) Detriments of performance in the traditional building process. *Journal of Construction Management and Economics* **10**: 343–359.

Mole, T. (1997) *Mind the Gap: An Education and Training Framework for Chartered Building Surveyors*. London, RICS.

Mooney, G.J., Gurbaxani, V. & Kraemer, L.K. (1995) A process oriented framework for assessing the business value of information technology. *Proceedings of the Sixteenth International Conference on Information Systems*, 10–13 December, Amsterdam, The Netherlands, pp. 17–28.

Muhegi, B. & Malongo, J. (2004) Globalisation: a challenge to developing countries. Paper presented at the *International Symposium on Globalisation and Construction*, 17–19 November, Bangkok, Thailand.

Mukhopadhyay, T. & Cooper, R.B. (1993) A microeconomic production assessment of the business value of management information systems, *Journal of Management Information Systems* **10**(1): 33–55.

Mullins, L.J. (1999) *Management and Organisational Behaviour* (5th edn). Harlow, Essex, Pearson Education.

Nam, C.H. & Tatum, C.B. (1988) Major characteristics of constructed products and resulting limitations of construction technology. *Construction Management and Economics* **6**: 133–148.

NAO 2001; Strategic forum for construction, 2002; Revaluing Construction, 2003.

Nason, T.W., Chamberlain, S., Rittase, W.M., Fleischer, W.R. & Morison, S.E. (1949) *Education Bricks and Mortar*. Cambridge, Mass., Harvard University.

Neff, T.L., Druss, D.L & Maswoswe, J.J. (1998) The STEPS Approach: A Realistic Method to Design and Construct Underground Facilities. *Tunneling and Underground Space Technology* **13**(2): 151–158.

NHS (2006a) from www.nhs-procure21.gov.uk.

NHS (2006b) *ProCure21 Elements–Facts and Figures*, from www.nhs-procure21.gov.uk.

NHS Estates (2000) *Sold on Health*. Leeds, NHS Estates.

Nicolas, B. (1999) Sector-specific paradigms: the dynamics of exemplary examples in organizations. Presented at the *Meso Organization Studies Group 10th Conference*, 30 April–2
May, Duke University, North Carolina.

Nkado, R.N. & Mbachu, J.I.C. (2002) Causes of, and solutions to client dissatisfaction
in the South African building industry: the clients' perspectives. Proceedings of the
First CIB W107 International Conference, 11–13 November, Stellenbosch, South Africa, pp.
349–368.

Nonaka, I. & Takeuchi, H. (1995) *The Knowledge Creating Company: How Japanese Companies
Create the Dynamics of Innovation*. New York, Oxford University Press.

Nowotny, H., Scott, P. & Gibbons, M. (2001) *Re-thinking Science: Knowledge and the Public in
an Age of Uncertainty*. Cambridge, UK, Polity Press.

NPWC/NBCC (1990) National Public Works Conference/National Building Construction
Council Joint Working Party, Canberra, Australia.

NSW (1992) *Royal Commission into Productivity in the Building Industry in New South Wales*.
Sydney, Australia, New South Wales Government, Vols 1–10.

Nutt, B. (1988) The strategic design of buildings. *Long Range Planning* **21**(4): 130–140.

Office of Science and Technology (1999) *A Better Quality of Life: A Sustainable Development for
the United Kingdom*. London, The Stationery Office.

Ofori, G. (1993a) Research in construction industry development at the crossroads. *Construction Management and Economics* **11**: 175–185.

Ofori, G. (1993b) Managing Construction Industry Development: Lessons from Singapore's
Experience. Singapore, Singapore University Press.

Ofori, G. (1994) Practice of construction industry development at the crossroads. *Habitat
International*, **18**: 41–56.

Ofori, G. (1996) International contractors and structural changes in host-country construction industries: Case of Singapore. *Engineering, Construction and Architectural Management*
3(4): 271–288.

Ofori, G. (2000a) Challenges for construction industries in developing countries. Proceedings of the *Second International Conference of the CIB Task Group 29*, November, Gaborone,
Botswana, pp. 1–11.

Ofori, G. (2000b) Globalisation and construction industry development: research opportunities. *Construction Management and Economics* **18**: 257–262.

Ofori, G. (2001) Challenges facing the construction industries of Southern Africa. Presented
at the *Regional Conference: Developing the Construction Industries of Southern Africa*, 23–25
April, Pretoria, South Africa.

Ofori, G. (2002) The knowledge-based economy and construction industries in developing
countries. Proceedings of the *First CIB W107 International Conference*, 11–13 November,
Stellenbosch, South Africa, pp. 419–428.

Ofori, G. (2003) Frameworks for analysing international construction. *Construction Management and Economics* **21**: 379–391.

Ofori, G. (2004a) Construction project risks from perspective of developing countries. Proceedings of the *Annual Conference 2004 of King's College London*, 1–2 July, Vol. 1, pp. 44–
59.

Ofori, G. (2004b) Revaluing construction in developing countries. *Mini-report for the CIB
World Building Congress 2004*. The Netherlands, CIB.

OGC (2006) *Achieving Excellence*. Office of Government Commerce (OGC),
www.ogc.gov.uk/guidance_achieving_excellence_in_construction.asp

Othman, A.A.E., Hassan, T.M. & Pasquire, C.L. (2004) Drivers for dynamic brief development in construction. *Engineering, Construction and Architectural Management* **11**(4): 248–258.

Oxford English Dictionary (1993) *The New Shorter Oxford English Dictionary: A–M*. Oxford, Oxford University Press.

Parker, M.M., & Benson, R.J. (1998) *Information Economics—Linking Business Performance to Information Technology*. New Jersey, Prentice Hall.

Pearce, D. (2003) The social and economic value of construction. In: *The Construction Industry's Contribution to Sustainable Development*. London, nCRISP.

Perspectives for Dutch Construction Sector, Policy Document of the Dutch Ministries of Trade and Industry, of Transport and Civil Works and of Housing, Spatial planning and the Environment (Perspectief voor de Bouw), 25 November 2003, The Hague.

Pietroforte, R. & Costantino, N. (2004) The construction engineering and management discipline in the USA: an analysis of two ASCE academic journals. Presented at the *CIB World Building Congress – Building for the Future*, 2–7 May, Toronto, Ontario, Canada.

Polanyi, M. (1964) *Personal Knowledge: Towards a Post-Critical Philosophy*. New York, Harper & Row.

Polanyi, M. (1966) *The Tacit Dimension*. Garden City, NY, Doubleday & Co.

Powell, J. (1995) *Towards a New Construction Culture*. Ascot, Chartered Institute of Building.

Preiser, W. (1985) *Programming the Built Environment* New York, Van Nostrand Reinhold.

Preiser, W.F.E. (2001) The evolution of post-occupancy evaluation: toward building performance and universal design evaluation. In: *Federal Facilities Council, Learning from our Buildings: A State-of-the-Practice Summary of Post-Occupancy Evaluation*. Washington, D.C., National Academies Press.

Preiser, W.F.E. & Schramm, U. (1997). Building performance evaluation. In: *Time-Saver Standards: Architectural Design Data* (ed. D. Watson, et al.). New York, McGraw-Hill.

Preiser, W.F.E. & Vischer, J.C. (2004). *Building Performance Assessment*. Oxford, Elsevier Science Publishers.

Preiser, W.F.E. & Vischer, J.C. (2005) *Assessing Building Performance*. Kidlington, Oxon, Elsevier.

Preiser, W.F.E., Rabinowitz, H.Z. & White, E.T. (1988) *Post-Occupancy Evaluation*. New York, Van Nostrand Reinhold.

PSIBouw (2004) *Dutch Construction Industry Reform Programme*. Rotterdam, PSIBouw, www.PSIBouwouw.nl

Raftery, J., Pasadilla, B., Chiang, Y.H., Hui, E.C.M. & Tang, B.S. (1998) Globalisation and construction industry development: implications of recent developments in the construction sector in Asia. *Construction Management and Economics* **16**: 729–737.

Remenyi, D. (2000) The elusive nature of delivering benefits from IT investment. *The Electronic Journal of Information Systems Evaluation* **3**(1), available online at www.itera.rug.nl/ejise/vol3/paper1.html.

Remenyi, D., Money, A. & Sherwood-Smith, M. (2000) *The Effective Measurement and Management of IT Costs and Benefits* (2nd edn). Oxford, Butterworth-Heinemann.

RIBA (1963) *Plan of Work for Design Team Operation*. London, RIBA Practice.

RIBA (1967) *Plan of Work*. London, RIBA.

RIBA (2003) The professionals' choice: the future of the built environment professionals. edited by F. S. London.

RICS (2004) *Electronic reverse auctions*. Coventry, RICS.

Rischmoller, L., Fisher, M., Fox, R. & Alarcon, L. (2000) 4D Planning and Scheduling (4D-PS): Grouping Construction IT Research in Industry Practice. In *Proceedings of CIB W78 Conference on Construction Information Technology: Taking the construction industry into the 21st century.* June, Iceland.

Roberts, P. (2004) FM: New urban and community alignments. Presented at *Futures in Property and Facility Management – Creating a Platform for Change*, 25–26 March, ICH Conference Centre, London.

Robbins, S.P. (1979) *Organisational Behaviour: Concepts, Controversies and Applications* (5th edn). New Jersey, Prentice-Hall International Editions.

Rowlinson, S.M. & Root, D. (1997) *The Impact of Culture on Project Management.* Final Report: Hong Kong/UK Joint Research Scheme, Department of Real Estate and Construction. University of Hong Kong.

Ruddock, L. (2000) An international survey of macroeconomic and market information on the construction sector: Issues of availability and reliability. *RICS Research Paper Series* 3(11). RICS, London.

Ruddock, L. & Lopes, J. (2006) The construction sector and economic development: The 'Bon' curve, *Construction Management and Economics* **24**: 717–723.

Ruddock, L. & Wharton, A. (2004) Revaluing construction, drivers from the urban environment – construction and sustainable development. *Mini-report for the CIB World Building Congress 2004,* The Netherlands, CIB.

Rwelamila, P.D., Talukhaba, A. & Kivaa, T.P. (2000) African intelligentsia – why have we embraced 'hyper barefoot empiricism; in procurement practices. In *Proceedings of the 2nd International Conference on Construction in Developing Countries,* 15–17 November, Gaborone, Botswana, pp. 457–466.

Sacks, J. (2003) *The Dignity of Difference* (2nd edn). London, Continuum.

Sanoff, H. (1977) *Methods of Architectural Programming.* Stroudsburg, Pennsylvania, Dowden, Hutchinson and Ross.

Scharle, P.M. (2002) Public–private partnership (PPP) as a social game. *Innovation* **15**(3): 227–252.

Schon, D.A. (1991) *The Reflective Practitioner.* New York, Ashgate Publishing Group.

Schumpeter, J. (1949) *Theory of Economic Development.* Massachusetts, Harvard University Press.

SCRI (2005) Constructing Futures on SCRI Strategy: Working Report. University of Salford, SCRI.

Senge, P.M. (1990) *The Fifth Discipline The Art & Practice of The Learning Organization.* London, Random House.

Senge, P.M., Scharmer, O.C., Jaworski, J. & Flowers, B.S. (2004) Awakening faith in an alternative future. *The SoL Journal on Knowledge, Learning and Change* **5**(7): 1–11.

Sexton, M. & Barrett, P.S. (2003a) Appropriate innovation in small construction firms. *Construction Management and Economics* **21**(6): 623–633.

Sexton, M. & Barrett, P.S. (2003b) A literature synthesis of innovation in small construction firms: insights, ambiguities and questions. *Construction Management and Economics* **21**(6): 613–622.

Shakantu, W., Zulu, S. & Matipa, W.M. (2002) Global drivers of change: Their implications for the Zambian construction industry. In: *Proceedings of the First CIB W107 International Conference,* 11–13 November, Stellenbosch, South Africa, pp. 133–143.

Shackle, G.L.S. (1971) *Economics for Pleasure*, Cambridge, Cambridge University Press.

Simon, E. (1944) *The Placing and Management of Building Contracts*. London, HMSO.

Simmonds, P. & Clark, J. (1999) UK Construction 2010: Future Trends and Issues Briefing Paper. London, CIRIA.

Singapore Manpower 21 Committee (1999) *Manpower 21: Vision of a Talent Capital*. Singapore, *Ministry of Manpower*. Available at http://www.mom.gov.sg/publish/momportal/en/home.html (accessed January 2004)

Slaughter, S.E. (1998) Models of construction innovation. *Journal of Construction Engineering and Management* **124**(3): 226–231.

Slaughter, S.E. (2004) DQI: the dynamics of design values and assessment. *Building Research & Information* **32**(3): 245–246.

SME Statistics (2002) *UK SME Statistics 2001*. Available at www.sbs.gov.uk (accessed May 2004).

Smith, J., Love, P.E.D. & Wyatt, R. (2001) To build or not to build? Assessing the strategic needs of construction industry clients and their stakeholders. *Structural Survey* **19**(2): 121–132.

Smith, R.A., Kersey, J. & Griffiths, A.J. (2002) *The Construction Industry Mass Balance: Resource Use, Wastes and Emissions*, Viridis Report No. 4 (Crowthorne: Viridis).

Smithson, S. & Hirschheim, R. (1998) Analysing information systems evaluation: Another look at an old problem. *European Journal of Information Systems* **7**: 158–174.

Snyman, G.J.J. (1999) Macroeconomic data for the South African construction industry: Historical development and new directions for research. In: *Macroeconomic Issues, Models and Methodologies for the Construction Sector* (ed. L. Ruddock). Rotterdam, CIB, publication number 240.

Soetanto, R., Dainty, A.R.J., Goodier, C.I., Harty, C.F., Austin, S.A., Price, A.D.F. & Thorpe, A. (2006) Synthesising emerging issues within key futures study reports in construction. Accepted for publication in: *Construction in the XXI century: Local and Global Challenges, Joint International Symposium of CIB Working Commissions W55, W65 and W86*, 18–20 October 2006, Rome.

Soh, C. & Markus, M.L. (1995) How IT creates business value: A process theory synthesis. *Proceedings of the Sixteenth International Conference on Information Systems*, 10–13 December 1995, Amsterdam, The Netherlands, pp. 29–42.

Spence, R., Macmillan, S. & Kirby, P. (ed.) (2001) *Interdisciplinary Design in Practice*. London, Thomas Telford.

SPTF (2006a) Sustainable Procurement Task Force. www.sustainable-development.gov.uk/government/task-forces/procurement/.

SPTF (2006b) Procuring the Future – Sustainable Procurement National Action Plan: Recommendations from the Sustainable Procurement Task Force. London, Department for Environment, Food and Rural Affairs, PB11710, downloaded from: www.sustainable-development.gov.uk/government/task-forces/procurement/

Staricoff, R.L. (2005) *Arts in Health: A Review of the Medical Literature. London*, Arts Council England.

Swan, W., McDermott, P., Cooper, R. & Wood, G. (2004) *Trust in Construction: Achieving Cultural Change* (ed. C.N. West). CCI, p. 21.

Szigeti, F. & Davis, G. (2002) Forum: The turning point for linking briefing and POE? *Building Research & Information* **30**(1): 47–53.

Szigeti, F. & Davis, G. (guest ed.) (2005) Special issue – performance-based building,. *Building Research & Information* **33**(2), Feb.

Szigeti, F., Davis, G. & Hammond, D. (2005) Introducing the ASTM facilities evaluation methodology. In: *Assessing Building Performance* (ed. W.F.E. Preiser & J.C. Vischer). Oxford, Elsevier, pp. 104–118.

Tallon, P.P., Kraemer, J.K. & Gurbaxani, V. (2001) *Executives' Perceptions of the Business Value of Information Technology: A Process-Oriented Approach, Working Paper #ITR-148*, 13 April 2001. University of California, Irvine, CA, Centre for Research on Information Technology and Organisations, Graduate School of Management.

Tan, W. (2002) Construction and economic development in selected LDCs: Past, present and future. *Construction Management and Economics* **20**(7): 593–599.

Tenner, E. (1996) *Why Things Bite Back: New Technology and the Revenge Effect*. London, Fourth Estate.

Ternouth, P. (2004) *The Business of Knowledge Transfer... Informing the Debate*. London, The Council for Industry and Higher Education.

The Straits Times (Singapore) (2004), Samy Vellu under siege over shoddy projects. October 16, p. A1.

Tilley, F. & Johnson, D. (1999) Modernization of universities through greater interaction with small firms. *Industry and Higher Education*, April.

Turin, D.A. (1969) *Construction and Development*. University College London, Built Environment Research Unit.

Turin, D.A. (1973) *The Construction Industry: Its Economic Significance and its Role in Development* (2nd edn). University College London, Building Economics Research Unit.

UNCHS (1991) *Technology in Human Settlements: The Role of Construction*. Nairobi, United Nations.

UNCHS (1995) *Policies and Measures for Small Contractor Development*. Nairobi, United Nations.

UNIDO (United Nations Industrial Development Organisation) (1969) *Industrialisation of Developing Countries: Problems and Prospects – Construction Industry*, Monograph No. 2. New York, United Nations.

Unison (2003) *LIFT: A Briefing for Non Experts*. Local Information Unit 2003, UNISON Stock No: 2235.

United Nations (2001) *International Recommendations for Construction Statistics*, Series M, No.47, Rev.1. New York, United Nations.

United Nations (2003) *Statistical Yearbook: 47th Issue*. New York, United Nations.

Valentin, E.M.M. (2000) University – industry co-operation: a framework of benefits and obstacles, *Industry and Higher Education*, June.

Van de Ven, A.H., Polley, D.E., Garud, R. & Venkataraman, S. (1999) *The Innovation Journey*. Oxford, Oxford University Press.

Vischer, J.C. (1989) *Environmental Quality in Offices*. New York, Van Nostrand Reinhold.

Vischer, J.C. (1993). Using feedback from occupants to monitor indoor air quality. *Proceedings IAQ93*, ASHRAE: Denver, June 1993.

Vischer, J.C. (1996). *Workspace Strategies: Environment as a Tool for Work*. New York, Chapman & Hall.

Vischer, J.C. (2001). Post-occupancy evaluation: a multi-facetted tool for building improvement. In: *Federal Facilities Council, Learning from our Buildings: A State-of-the-Practice Summary of Post-Occupancy Evaluation*. Washington, D.C., National Academies Press.

Vischer, J.C. (2005) *Space Meets Status: Designing Workplace Performance.* London, Routledge.

Von Foerster, H. (1985) *Epistemology and Cybernetics: Review and Preview.* Lecture held on 18 February at Casa della Cultura, Milan, Italy.

VROM (2004) *International Clients Platform – Meeting Report.* Hoftoren, The Hague, CIB, p. 11.

VTT (ed.) (2005) *Pre-Summit Study: Global Synthesis Report.* VTT, p. 10.

Van Wagenberg, A.F. (1997) Facility management as a profession and academic field. *International Journal of Facilities Management* **1**(1): 3–10.

Walker, D. & Hampson, K. (2003) Preface. In: *Procurement Strategies: A Relationship-based Approach* (ed. D. Walker & K. Hampson). Oxford, Blackwell Science Ltd.

Walsham, G. (1993) *Interpreting Information Systems in Organisations.* Chichester, John Wiley.

Wells, J. (1986) *Construction Industry in Developing Countries: Alternative Strategies for Development.* London, Croom Helm.

Wells, J. (1987) *The Construction Industry in Developing Countries: Alternative Strategies for Development.* London, Croom Helm.

Willcocks, L. & Lester, S. (1994) Evaluating the feasibility of information systems investments: Recent UK evidence and new approaches. In: *Information Management – The Evaluation of Information Systems Investments* (ed. L. Willcocks). London, Chapman & Hall.

Williams, A. (2004) Industry engagement in UK built environment higher education courses. *ARCOM Conference*, September 2004, Heriot-Watt University, Edinburgh.

Winch, G.M. (2003a) Models of manufacturing and the construction process: the genesis of re-engineering construction. *Building Research & Information* **31**(2):107–118.

Winch, G.M. (2003b) How innovative is construction? Comparing aggregate data on construction innovation and other sectors – a case of apples and pears. *Construction Management and Economics* **21**(6): 651–654.

Winch, G.M., Courtney, R. & Allen, S. (2003) Editorial: re-valuing construction. *Building Research and Information*, **31**(2), 82–84.

World Bank (1984) *The Construction Industry: Issues and Strategies in Developing Countries.* International Bank for Reconstruction and Development. Washington, D.C., The World Bank.

World Bank (2004a) *World Development Indicators.* New York, World Bank.

World Bank (2004b) *World Development Report, 2005: A Better Investment Climate for Everyone.* Washington, D.C., World Bank and Oxford University Press.

World Bank (2005) *World Development Indicators.* www.worldbank/org/data/wdi2005

Woudhuysen, J. & Abley, I. (2004) *Why is Construction so Backward?* Chichester, John Wiley.

Zaghloul, R. & Hartman, F. (2003) Construction contracts: the cost of mistrust. *International Journal of Project Management* **21**: 419–424.

Zeisel, J. (1984) *Inquiry by Design.* Cambridge, Cambridge University Press.

Zeisel, J. Silverstein, N.M., Hyde, J., Levkoff, S., Lawton, P.M. & Holmes, W. (2003) Environmental correlates to behavioural health outcomes in Alzheimer's special care units. *The Gerontologist* **43**(5): 697–711.

Ziman, J. (1994) *Prometheus Bound: Science in the Steady State.* Cambridge, Cambridge University Press.

Other sources

Department of Trade and Industry: www.dti.gov.uk
HESA – Higher Education Statistics Agency: www.HESA.ac.uk
Lees, M. & Lungu, C. (2004) A study of graduate experience – project ongoing (unpublished)
UCAS – Universities and Colleges Admissions Service: www.UCAS.ac.uk
Websters New World Dictionary (1956). London, MacMillan.

Index

4Cs model, 197
A vision, 177
Action areas, 217
Action learning, 50
Actions, 210
Advanced countries, 69
Alliances, 44
Ambition, 196

Banking systems, 182
Benchmarking, 34
Benefits, 52
Best
    experiences, 206
    practice, 48
    value, 112
Briefing, 42, 149, 152, 155, 205
Building life-cycle, 13, 18, 52
Building performance evaluation (BPE),
    39, 51, 83, 155
Built wealth, 79
Business
    case, 147
    cycle, 85, 104
    value, 123, 127

Capacity building, 189
Capital stock, 76, 176
Catalyst, 54
Championing, 191
Change, 163
Cheapest, 152
Clients, 25, 46, 58, 83, 88, 90, 106, 186, 189,
    206, 215
Clients' association, 21, 93
Clusters, 15
Coalition, 172
Codes of Ethics, 89
Collaboration, 104, 196, 206
Colonial, 186

Comfort, 159
Communication, 119, 139, 155
Community, 196
Company-based improvements, 173
Competence, 142
Competition, 32, 84, 85
Competition policy, 83, 90, 102
Competitiveness, 51, 101, 107
Confidence, 217
Constraint, 196, 208
Construction
    change agent, 11
    reports, 107
    workforce, 44
Construction Industry Development
    Board (CIDB), 179
Continuing pressure, 101
Continuing professional development, 86,
    190
Contribution, 68
Cooperation, 85, 185
Co-ordinated, 120
Cost, 151
Covenantal relationships, 31
Creating value for all, 28
Creativity, 196
Cultural change, 154
Culture, 86, 103, 132, 186, 187
Curriculum, 131
Cycle, 117

Data objects, 117
Decision-making, 152, 156
Demand for recruits, 133
Demographic trends, 55
Deregulation, 183
Design, 20, 62, 155
Design quality indicators, 40
De-skilling, 142
Developing countries, 69, 132, 173, 174, 175

Diagnostic measuring, 160
Dialogue, 218
Differentiating, 171
Differentiation, 170
Disjointed processes, 45
Distribution of rewards, 27
Downstream players, 43
Drivers, 210
Dutch experience, 87
Dynamic balance, 97, 164, 206
Dynamism, 62

Economic growth, 176
Economic measures, 67, 84
Education, 121, 130, 139, 190
Education research, 215
Education supply chain, 132
Effective practitioners, 131
Effectiveness, 124, 126
Efficiency, 126
Elegant solutions, 20, 52
Employment, 168
End-users, 93
Energy consumption, 81
Energy performance, 56
Engagement, 144, 147
Environmental issues, 56
Environmental quality, 152
Environment–behaviour links, 52
Exemplars, 195

Facilities management, 74, 152
Feedback, 154
Financing, 150
Fitness for purpose, 92
Fixed capital formation, 76
Flexibility, 62, 106, 208
Focus, 218
Fragmentation, 62
Frameworks, 209
Free market, 110
Functional comfort, 161

GDP, 13
Globalisation, 183
Goal, 153, 163
Good practice, 49

Government, 108, 188, 210, 213
GVA, 60, 168

Habitable environments, 56
Hotspot, 135
Housing needs, 178

Iconic, 62
Iconic structures, 53, 62
ICTs, 38
Image, 3, 52, 87, 88, 137, 148, 181
Impacts, 195, 216
Implementation, 25, 120, 186
Improvement, 58
Incentives, 100
Industry, 3
   client, 210, 212
   dynamics, 86
   reform, 86
   reviews, 176
Industry Foundation Classes (IFCs), 118
Infinity model, 7, 218
Information flow, 37
Information technology (IT), 115, 215
Innovation, 62, 89, 106, 150, 169, 215
Instinct, 129
Institutions, 216
Integrated, 116, 146, 215
Integration, 36, 99, 129, 171, 182
Interactive effects, 52
Interconnectivities, 106
   interactive effects, 52
   interdependency, 103
   inter-disciplinary, 147
Interest group, 217
Interoperability, 36
Interventions, 110
Investment, 160

Joint ventures, 185

Kaleidoscopic model, 49
Knowledge, 140, 161
   attitudes, 209
   management, 48

Leading change, 205

Lean construction, 43, 95
Learning, 86, 131
Least cost, 206
Legal/tax mechanisms, 213
Liaison, 203
Life-building, 117
Lifelong learning, 87, 144
Lifestyle, 137
Local firms, 184
Location specificity, 182
Lock-in, 206
Long-term, 84
  process, 102
  relationships, 33
  strategy, 103
Looking in, 7, 218
Looking out, 8, 218
Lowest cost contracts, 30
Lowest price, 102
  mentality, 48
Loyal workforce, 48

Maintenance, 217
Management, 215
Managerial capability, 184
Market
  dynamics, 84
  structures, 87
Measures of success, 206
Meso-economic analysis, 11
Minimum standards, 51
Mode 1, 145
Mode 2, 145
Momentum, 101
Multiple perspectives, 125

National development, 176
nD CAD, 38, 115
Network organisations, 95
New construction, 75
New professions, 151
Nodes for development, 213
Non-linear process, 42
Non-price factors, 90, 103

Operating budgets, 153
Organisational learning, 38

Paradigm, 84, 145, 208
Partnerships, 32, 94, 96, 113, 202, 286
People, 149
Perceived performance, 159
Performance, 3, 7, 127, 158, 167, 181
  based building, 39, 93, 97
  criteria 5Es, 17, 20
  data, 91, 122
  evaluation, 158
  specifications, 44
Physical comfort, 161
Pleasure, 207
Points of risk, 98
Policies, 188
Political push, 100
Pollution, 81
Positive
  image, 62
  view, 58, 164
Post occupancy evaluation (POE), 153
Postal survey, 5
Poverty alleviation, 180
Practice, 49, 177, 186
Pre-design, 149
Pre-qualification, 33, 112
Price, 96
Prisoner's dilemma, 30
Private sector, 92
Privatisation, 112
Problem-based learning, 141
Problem-solving, 171
Process of reform, 23
Procurement, 91, 93, 105, 107, 123, 188, 214
Professional bodies, 25, 33, 89, 142, 178,
  190
Professionalism, 33, 87, 178, 207
Project, 209
  team, 214
Promotion, 62
PSIBouw, 84
Psychological comfort, 161
Public sector, 90, 101
Purpose, 172

Quality, 95, 162
  of life, 51, 67, 81

Rational process, 43
Recycling, 158
Re-engineering, 43
Reform, 83, 90
Regeneration, 54
Relationship model, 59
Repair and maintenance, 71, 74
Research, 49, 215, 221
Restrainers, 210
Return on investment, 92
Re-use, 158
Revaluing construction, 4
Risk management, 34
Role of government, 83

Satisfaction, 159, 207
Scale of activity, 51
Scandals, 88, 90
Scenarios, 114
Self-employment, 165
Self-regulation, 87
Self-sustaining changes, 24, 101
Service orientation, 35, 71, 109, 123
Simulate, 117
Small and medium-sized enterprises SMEs, 69, 166, 172
Small firms, 62
Social
  capital, 31
  change, 54
  cohesion, 182
  costs, 150
Society, 216
Socioeconomic development, 178
Specialisation, 62
Stakeholders, 13, 20, 21, 87, 120, 155, 189, 209, 212, 216
Standard Industrial Classification (SIC), 11, 62, 68
Statistical information, 82
Strategic alliances, 182
Strategy, 87, 156, 177
Structure, 69
Sub-clusterings, 15

Success, 208
Supply of recruits, 136
Sustainable construction, 180
Sustainable development, 57, 67, 200
Synergies, 52
Synthesis map, 6
Systemic impacts, 57, 114
Systems model, 156

Tacit knowledge, 41
Targets, 177, 199
Task environment, 171
Team dynamics, 96
Technical correctness, 59
Technology, 41, 146
Temporary coalition, 14
Territory, 161
Trade-offs, 118
Training/education, 139, 215
Transaction cost, 39, 59
Transformation, 106, 221
Trust, 32, 42, 44, 84, 85, 87, 89, 186
Turbulence, 62, 169

Understanding, 49
Urban FM, 54
Urbanisation, 79
Use-phase, 40
Users, 149, 150, 160, 189, 216

Value
  adding, 68
  adding activities, 18
  management, 44
  streams, 57
  versus costs, 92
Virtuous cycle, 218
Visibility, 151
Vision, 22, 84, 87, 103, 213
Vocational education, 138, 141

Waste, 116
Wealth, 68
Whole life-cycle, 74, 120
Work environments, 150